Sampling for Gold

by Arizona Bureau of Mines

Edited by Kerby Jackson

Introduction

It has been decades since the Arizona Bureau of Mines released these important publications on sampling for gold. First released in 1918, these important volumes have been out of print and has been unavailable to the mining community since those days, with the exception of expensive original collector's copies and poorly produced digital editions. They are compiled here as a group for the benefit of today's prospector.

It has often been said that "*gold is where you find it*", but even beginning prospectors understand that their chances for finding something of value in the earth or in the streams of the Golden West are dramatically increased by going back to those places where gold and other minerals were once mined by our forerunners. Despite this, much of the contemporary information on local mining history that is currently available is mostly a result of mere local folklore and persistent rumors of major strikes, the details and facts of which, have long been distorted. Long gone are the old timers and with them, the days of first hand knowledge of the mines of the area and how they operated. Also long gone are most of their notes, their assay reports, their mine maps and personal scrapbooks, along with most of the surveys and reports that were performed for them by private and government geologists. Even published books such as this one are often retired to the local landfill or backyard burn pile by the descendents of those old timers and disappear at an alarming rate. Despite the fact that we live in the so-called "Information Age" where information is supposedly only the push of a button on a keyboard away, true insight into mining properties remains illusive and hard to come by, even to those of us who seek out this sort of information as if our lives depend upon it. Without this type of information readily available to the average independent miner, there is little hope that our metal mining industry will ever recover.

This important volume and others like it, are being presented in their entirety again, in the hope that the average prospector will no longer stumble through the overgrown hills and the tailing strewn creeks without being well informed enough to have a chance to succeed at his ventures.

Kerby Jackson
Josephine County, Oregon
May 2016

SAMPLING AND THE ESTIMATION OF THE GOLD IN A PLACER DEPOSIT

BY GEO. R. FANSETT.

The sampling and the estimation of the gold present and recoverable play a very important part in the history and development of all placer deposits. For this reason everyone interested in placer mining should know the best methods of working this class of mineral deposit. The purpose of this bulletin is to indicate and describe to the miner and to the layman the best general methods used, so that they may understand and know how to do each and every part of the work, as well as to make the necessary calculations.

The work may be divided into three general subdivisions, namely: (1) The Sampling of the Deposit, (2) The Testing of the Samples, and (3) The Estimation of the Total Gold Present and the Gold Recoverable in the Deposit.

A placer is a deposit of mineral-bearing gravel, sand or soil. The commonest forms referred to are gold placers, tin placers, and platinum placers. The same method of procedure can be used in any kind of placer, but this bulletin particularly refers to gold placers, as they are the most common and of the most importance in the United States.

A sample is a collection of fragments or pieces from a deposit which contains exactly the same minerals in exactly the same proportions as they exist in the deposit. The act of collecting these pieces or fragments is called sampling.

The gold present is the amount of gold actually existing or present in the deposit. Gold is sold at the rate of $20.67 per ounce Troy, but results from assayers are calculated at the rate of $20 per ounce Troy. The ton used is the short ton of 2000 pounds avoirdupois. The value of silver fluctuates so that no definite value is given.

The gold recoverable is the amount of gold which can be extracted from the deposit by the use of any of the well known processes, such as a pan, sluice box, concentrator, centrifugal separator, dry washer, or any other process of ore dressing. There are always losses in this work, and for this reason the gold actually recovered is always less than is present in the deposit.

Sampling is slow work, and the greatest care is absolutely necessary and must be used in every detail of the work. Slipshod, careless work is absolutely out of the question, because if any one part of the work is not properly done, the collection of fragments obtained will not be a true sample of the deposit, and is not only valueless for the purpose intended, but will result in an incorrect estimate of the deposit.

RECORDS.

In the sampling of placers it is necessary to keep a good set of records, so that any questions which may arise can be correctly answered and definitely settled. For this purpose a topographical map, a sample book, and a diary are used to record all matters connected with the sampling of the deposit. These must be very carefully kept, as from one or all of them important points have to be decided.

If there arises a question in one's mind as to where to record a certain detail which seems to belong in two of the records, record it in both of them; and if it seems to apply to all of the records, record it in all of them, and do not consider it useless repetition—any matter worth recording is worth finding easily.

THE TOPOGRAPHICAL MAP.

A topographical map is made of the entire deposit as soon as a thorough preliminary examination has been completed. This map should be made to a scale large enough so that all details can be plainly marked.. All distances for this map are measured on the horizontal. All elevations are calculated from a permanently fixed benchmark or datum.

When the location of a test pit or hole is decided upon and work started, the location, together with the number given to it, is marked on the map. Likewise, when a test hole or pit has reached bedrock, the elevation of the bedrock is recorded. On the map is also recorded, by their numbers, the location of each of the samples, and, after they have been assayed, the values obtained. This map is used to record everything of this nature concerning the deposit, and should be so well kept that its records, together with those kept in the sample book and in the diary, will answer all inquiries which may come up.

THE SAMPLE BOOK.

Supplementary to the topographical map a sample book should be kept. In this book all records and details pertaining specifically to each sample are recorded, such as the name of the deposit, number of the sample, its location, the date when taken, kind of soil, assay returns, etc., etc.

A convenient form of sample book is one which has a page which is perforated near the far end so that this end part or tag can be easily torn off. The number given the sample is written on the part of the page which remains in the book and also on the detachable part. The sample book should be kept so that the information in it, together with the information on the topographical map and in the diary, will answer all possible questions which may come up regarding the deposit.

The form on following page represents one of the pages used in a sample book, but this form should be changed if necessary to meet the requirements of any particular job.

DIARY.

Together with above records a diary should be kept of the deposit. All matters of importance which are not naturally included among those recorded on the topographical map or in the sample book are taken care of in the diary. Such matters as the date when work started on a test pit, when and to what extent a certain pit caved, etc., should be recorded in the diary.

NUMBERING OF SAMPLES.

Each sample is given a different number, irrespective of its location, and by this number the sample is identified at all times. When the sample is taken the number is marked on the map at the proper location from which the sample was taken. The number is also marked on a page in the sample book and on the detachable part of the page or tag of the sample book, if paper tags from the sample book are used. When the sample has been cut down to the desired size it is put into a sample sack or container, together with the tag bearing its number.

TAGS.

Only one tag is put into each sack with each sample. The tags used for this purpose are made either of paper, soft wood, or metal. Paper tags are usually used except when the samples are very wet. These are usually the detachable part of the page referred to in connection with the sample book, but can also be any piece of paper with the proper number written on it. These are rolled up tight in the form of a lead pencil and have their ends well crimped. This is

TAG

THIS PART REMAINS IN SAMPLE BOOK

SAMPLE NO.

19 _____ SAMPLE NO. _____

NAME OF DEPOSIT _____

HOLE NO. _____

KIND & CONDITION OF DIRT _____

WEIGHT OF DIRT PER CU. YD. _____

METALS

ASSAYS	GOLD.	OUNCES PER CU. YD	VALUE PER CU. YD		

REMARKS

BLANK PAGE FOR A SAMPLE BOOK

done so that they will not unroll and get soiled, thus keeping the writing legible

Metallic tags having the number stamped on them are sometimes put into the sacks with the sample insttead of the paper tags. Soft wooden tags with the numbers written on them with a hard lead pencil are also convenient. A hard pencil is recommended for writing the numbers on the wooden tags because it will cut into the wood and even if the lead is rubbed off the indentation will be left. This kind of tag is very useful with wet samples.

Where any other tag is used instead of the detachable tag of the sample book, all the information in regard to the sample should be recorded in the sample book in the same manner as was explained above.

SAMPLE SACKS.

Only new sample sacks should be used. Used sacks may contain values from the former samples which they have held, which values may get mixed with the sample, thus enriching or salting it. It is poor business to spend a large sum of money on a collection of fragments from a deposit, in obtaining the sample, testing, shipping, and other charges, if the collection is not a true sample of the deposit. New sacks cost only a few cents each and should always be used. If tin containers are used they must be thoroughly washed and cleansed before the sample is put into them.

SAMPLING PLACERS BY THE USE OF TEST PITS.

One of the common methods used for sampling placers is in the use of test pits. The method of procedure is the same whether the hole is a shaft, pit, drift, raise, or other form of excavation. The one point to be kept in mind is that *all* of the material excavated from the hole decided upon is the sample.

Test pits can be used to advantage where the ground is stable enough to stand up well, where the deposit is not too deep or where water does not interfere too much.

The sample from a test pit is considered by many engineers to be better, in some respects, than the samples taken from drill holes. This is based on several assumptions;

(1) There is not the tendency for the values to concentrate in the lower level or bottom of the pit if the work is properly performed.

(2) Sectional samples or samples for any particular depth at a given point can easily be taken.

(3) The bulk of the sample from a test pit is much greater than that from a drill hole. For this reason, if small additions of values take place they will not affect the final result to so great an extent.

(4) In the test pit it is possible to see clearly the formations, and thus a better judgment of the deposit can be formed.

LOCATION AND NUMBER OF TEST PITS.

When test pits are used, great care and judgment should be exercised in locating the pits so as to get a fair final sample. Usually one pit for each two or three acres is sufficient. This is the number which was found ample in the sampling of several placers, but experience and judgment alone can decide this exceedingly important point for any particular placer deposit. As a general rule, the richer the deposit, the more pits are necessary. Also the greater number of pits are needed where the depth of the deposit is the greatest, as this represents the greater part of the tonnage. The sides or upper edges of the deposit should not be overlooked as these serve as indications as to whether the placer becomes richer or poorer as one works toward the higher benches.

FORM OF PIT.

The pits are usually rectangular in form, about three feet wide, and long enough so that a man can work to advantage. If the excavation is in the form of a drift, it must, of course, be high enough to work in comfortably. The sides should be as nearly perpendicular as possible. If the ground caves, it is necessary to timber it. If this happens, much care must be taken that none of the material outside of the section of the pit decided upon is included in the sample.

CAREFULNESS IN THE HANDLING OF THE SAMPLE.

The greatest care must be taken in handing the material. This is especially true after it has been taken out of the excavation. The safe way is to shovel or dump it directly onto a tight platform. If a platform is not available, a large sheet of steel or a large piece of thick canvas will answer the purpose. This is done so that no foreign material will get mixed with the sample. Since the bulk of the sample is usually much greater than is needed for the tests to be run (50 lbs. is usually more than is needed for panning and assaying), and too much to be handled conveniently, the next step is to cut or quarter it down to the size desired.

METHODS USED TO CUT DOWN THE SAMPLE.

Where mechanical quartering machines are not available and the sample is over 800 pounds, all of the following methods can be used, in their order, to advantage in quartering down the sample. When the sample is less than 500 pounds and over 100 pounds, the Cornish method of quartering can be used. When the sample is less than 100 pound, the method using canvas or oilcloth is good.

CUTTING DOWN SAMPLE WHICH WEIGHS OVER 800 POUNDS.
SHOVEL METHOD.

The first of the cutting down to about 500 pounds can be done by the use of shovels. In this method every second shovelful is passed to another platform or a different part of the same platform, which has been well cleansed. The odd ones are discarded. This operation cuts the sample in half. If it is still much too large (over 800 pounds) the above operation is repeated until the sample gets to between 300 pounds and 800 pounds. After it is cut down to this size (about 500 pounds) it is better practice to use the Cornish or Coning method down to about 100 pounds.

CORNISH OR CONING METHOD OF QUARTERING.

When the sample is not much over 700 pounds this method is used conveniently. The last heap of material from the preceding work of cutting down is leveled to a circular form not over four inches deep, by the use of a hoe, flat-nosed shovel, or a similar tool. The next step is to cone it, which is done in the following manner: From the outside of this leveled heap, at points equally distant from each other, equal amounts are shoveled up and allowed to fall onto the center of the

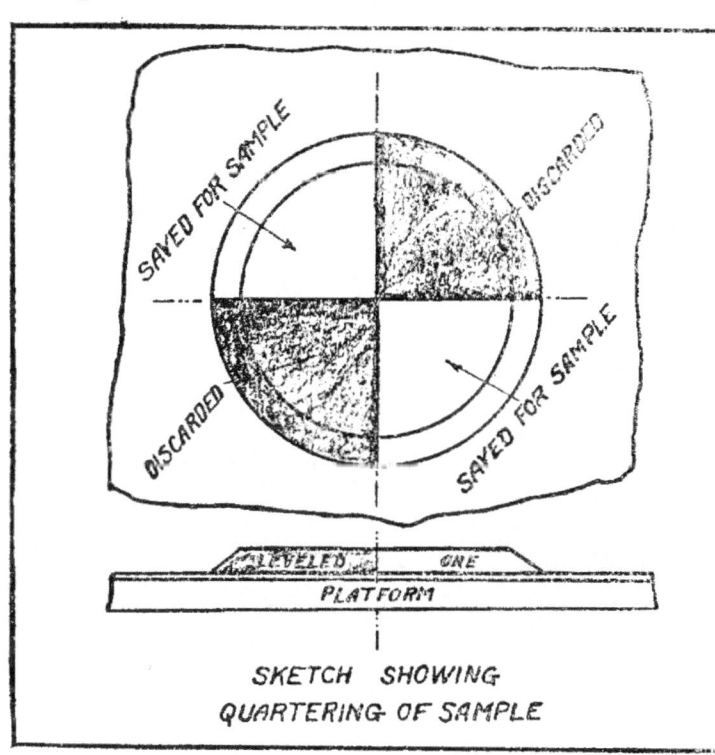

SKETCH SHOWING
QUARTERING OF SAMPLE

leveled heap in such a manner that the material is evenly distributed on all sides of the cone which is formed. In this way only a portion of the heap is shoveled up is passing once around it, thus making an even distribution of the values.

After all of the material outside of the cone is piled up in this manner onto the cone, the material is removed to a different platform or a clean part of the same platform and leveled and coned again. This process is repeated until the material has been thoroughly mixed. When this has been accomplished it is then leveled again to a circular form not over four inches deep, and divided into equal quarters by cutting it along two diameters at right angles to each other. The two opposite quarters are discarded.

Much care must be taken in doing this work, as well as to clean thoroughly the parts of the platform where the discarded metal has been in order to prevent salting. If the sample is still too large, more than 200 lbs., the two quarters which were left from the preceding operation are removed to a clean part of the platform where it is leveled, coned, and leveled again to a circular form and divided as before, the two

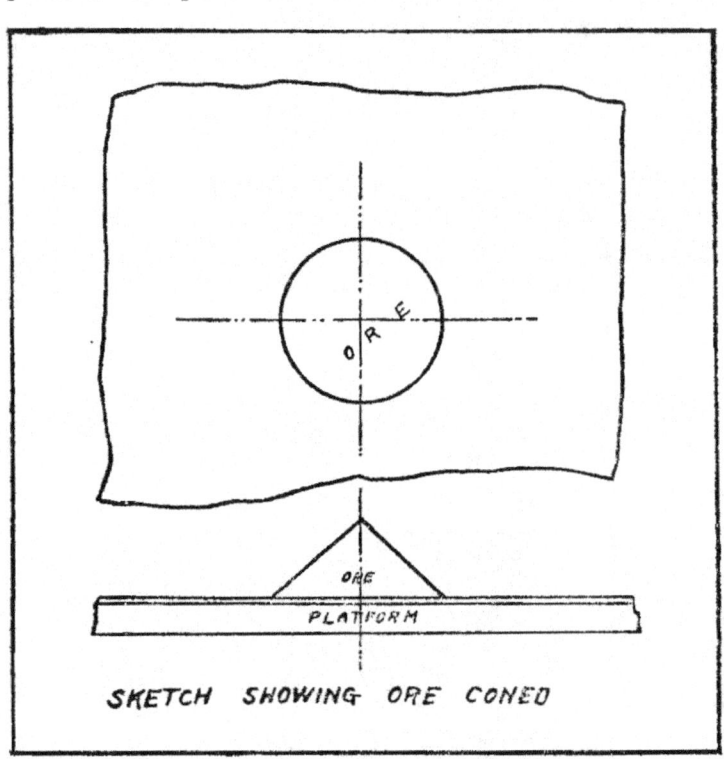

SKETCH SHOWING ORE CONED

quarters being again discarded. This operation is repeated until the sample reaches about 100 pounds, when the following method should be used:

QUARTERING BY USE OF CANVAS OR TABLE OILCLOTH.

The final sample from the preceding work is shoveled onto a square piece of table oilcloth or canvas about 6 feet by 6 feet. After this is done the two opposite corners of the oilcloth are taken, one in each hand. While one of the corners is slowly lowered, the other is raised at the same rate, the lower part of the oilcloth always resting on the ground or platform. This motion rolls and mixes the sample. After this is well performed the first two corners are allowed to fall flat, while the other two are taken in the same way. The sample is again rolled and mixed, but in the opposite direction to the first mixing. This is repeated several times until the sample is thoroughly mixed.

When this has been accomplished satisfactorily the canvas is spread out flat and the sample is leveled to a circular form as described before in the coning method. It is then divided into equal quarters by cutting it along two diameters at right angles to each other, as illustrated in figure (2). The two opposite quarters are discarded as in the coning method and the space where they were is thoroughly cleansed. If what remains is still too large the entire operation is repeated until the sample reaches the desired size. After thoroughly mixing it again the sample is put into a sack with the tag bearing its number for identifying it. It is then ready to be tested.

SECTIONAL SAMPLES.

These are samples taken at a given depth to indicate how the values run at that particular depth. They are usually taken from a section which has the same width completely around the pit or drift. When it is necessary to timber, these samples must be taken before the timber is put in place.

In the sample book record the number of the sample, the date when taken, the number of the pit, and all other data bearing on the sample. Measure down to the top of the section to be sampled from a fixed known elevation, such as the permanent ground level or a nail or stake solidly driven in at a definite elevation. Record this measurement in the sample book, together with the width of the section to be sampled.

The sample is the material which is picked off evenly from every part of this section. A sample pick, drill, or similar tool can be used. A box or other receptacle is held below so that all the material will be caught as it is broken off. A powder box answers this purpose very well. It is well to spread a piece of canvas on the ground below the place from which the sample is taken so as to catch any pieces that may not fall into the box. If the sample is too large it can be cut down, by one of the methods explained before, to the desired size. After being thoroughly mixed, the sample is put into the sack with the tag bearing the proper number, and is then ready to be tested.

DRILL SAMPLING.

Sampling by drilling is done either by the use of an augur drill, or by a churn drill outfit.

AUGUR DRILL METHOD.

The augur drill is a tool similar to that used in drilling post holes. The hole is made by the augur as it is turned, the material raised on the blades of the augur being the sample. It can only be used to advantage in soft ground which does not cave too badly, and only to a limited depth. The caving may be overcome by casing the hole.

Sampling by this method, where it can be used, is by far the cheapest and quickest of any method. It has the disadvantage, in some cases, of having the values concentrate at the bottom of the hole, from which it is next to impossible to extract them, thus tending to give an unfair sample.

The same care and experience in locating the holes to be drilled must be used in this method as in the location of the test pits. The records for sampling by this method are kept in the same manner as for test pits, and the same method for cutting down the samples can be used. After they have been cut down to the desired size they are well mixed, put into a sack with the tag bearing the proper number, and are ready to be tested. The testing of the samples and the method of calculation used will be treated later.

SAMPLING PLACERS BY THE USE OF A CHURN DRILL OUTFIT.

This is one of the very few methods which can be used to advantage when the deposit is deep, where the ground caves and where water interferes. For deep holes the cost per foot of hole drilled is low and the work usually progresses very rapidly.

There is the disadvantage of the possible concentration of the values in the bottom of the hole, from which it is difficult to extract them. The values from around the sides of the hole may also be washed down into the sample and salt it. Sectional samples by this method are not considered to be very satisfactory. An expensive outfit is needed to do the work by this method, and unless much drilling is to be done it is found better to let the job to a reliable churn drill contractor who will furnish the rig, casing, and other supplies needed, as well as the skilled operators who are absolutely essential for doing the work satisfactorily. Where the ground caves or water interferes it is necessary to case the hole.

LOCATION OF THE HOLES TO BE DRILLED.

The same care and judgment should be taken in the locating of the points where the holes are to be drilled as is necessary in the location of test pits, and the same matters should be borne in mind when doing this work.

The same general method of procedure in mapping the deposit, taking and numbering the samples, and the care necessary in handling them, must be used for the churn drill work.

The material taken from the hole is usually crushed by the operation of the drill bit to a size smaller than a walnut. If, as in most cases, the sample is much larger than is needed for testing, it can be cut down to the desired size by using the methods heretofore described. The sample from the churn drill hole seldom contains pieces much larger than a walnut and can be cut down very rapidly and accurately by the use of a Jones Sampler. This is an inexpensive apparatus which stands rough handling and is not very bulky. If the work is properly done, the final cut down sample is absolutely accurate and for this reason this machine can be used to advantage, providing the pieces in the sampler are not over one-half as big as the width of the slot in the sampler which is used. It is sometimes advisable to use one large sized sampler in the first part of the cutting down work, cutting down to the final size by using a smaller sampler, the slots of which are narrower than than in the first.

JONES SAMPLER.

On top, the Jones Sampler has a row of horizontal slots, all of which have the same length and width. From each one of the slots runs a chute, every second chute running in the same direction; that is, the outlet from the first chute and slot is opposite to the outlet from the second chute and slot. The material is shoveled slowly from a flat-nosed shovel or scoop transversely onto these slots, much care being taken that they do not clog, the material distributed evenly by moving the shovel back and forth from one side to the other of the machine. The end of the shovel should be held about one inch above the slots of the sampler and in this way equal amounts of the sample will fall into each slot and run down and out of the chute under it.

A pan is placed under each row of chutes to catch the material. If the work is properly done, exactly one-half of the material will be caught in each of the pans, and each pan will contain the same values. One operation of the sampler cuts the sample in half. The material in one of the pans is saved for the sample, while that in the other is thrown away. If the sample is still too large, the above operation is repeated until it reaches the desired size. It is then put into its sack with the tag bearing the proper number, tied up, and is ready to be

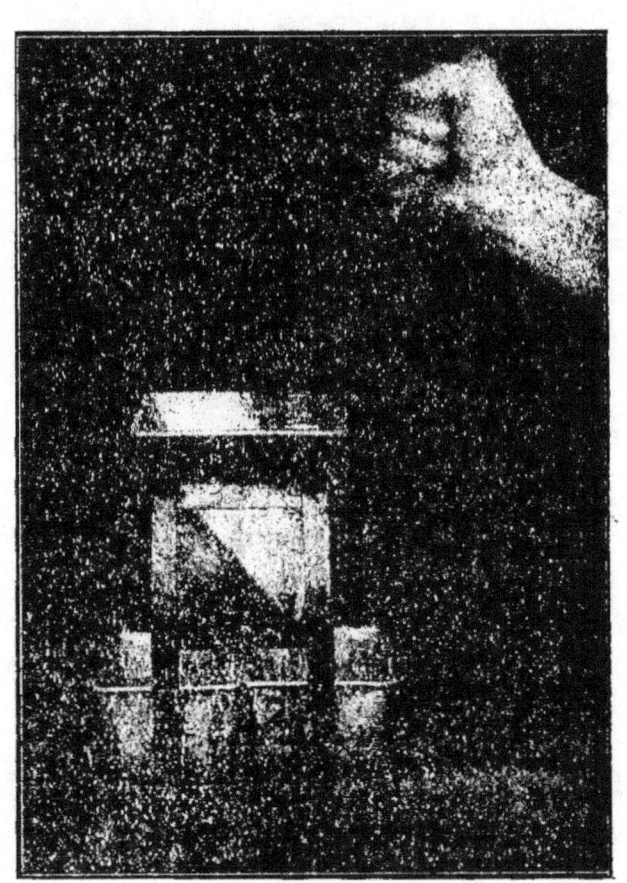

tested.

TESTING THE SAMPLES—PURPOSES OF THE TESTING.

The amounts of valuable mineral present and recoverable are the most important points to be decided from the test. Supplementary to this are several other important matters which should be taken into consideration, such as the form in which gold occurs, whether fine, coarse, or flaky; whether clay is present, and to what extent, and all other points which will determine whether or not the deposit can be worked at a profit.

METHODS OF TESTING.

Samples from gold placers are tested by panning or by the use of the dry washer or concentrator, or by both. Only those deposits where the dry washer can be used for working the deposit should be tested by the dry washer. These placers are only found in the deserts, where water for washing the ground is out of the question and where the ground is absolutely dry and free from clay. In all other placers the samples are usually tested by panning. After the panning tests prove satisfactory, tests on a large scale are usually made by using the method decided upon for doing the work.

For testing samples by panning the most important thing is a thoroughly experienced panner. The tools needed are good gold pans,

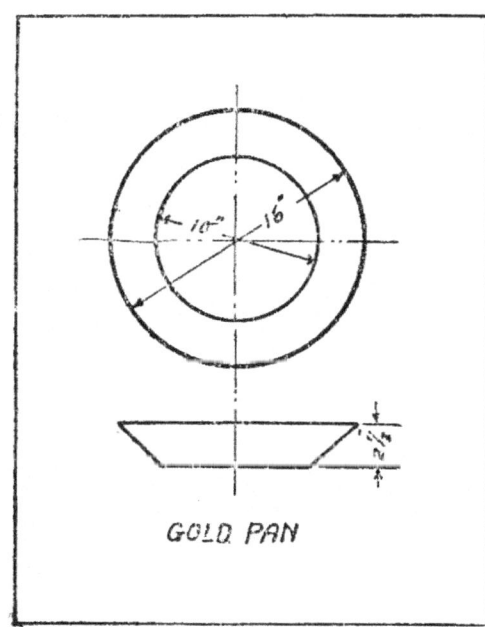

GOLD PAN

a measuring box, small pans, glass vials, a good pair of balances, and tags. The pans used are the regulation gold pans made from sheet iron and must be absolutely free from grease and rust. They are usually 16 inches in diameter on top, 10 inches in diameter on the bottom, and 2½ inches high.

When the value of the gold obtained from the panning of one or more boxes of the dirt is known, it is an easy matter to calculate the value for a cubic foot or cubic yard of the dirt. The following example will illustrate this point:

Suppose four boxes of dirt were panned, and from them were recovered $0.02 of gold. (Gold figured at $20 per ounce Troy.) Since there are 8 of the 6″x6″x6″ boxes in one cubic foot, and there are 27 cubic feet in one cubic yard, the calculation for this case is the following:

$0.02× 2=$0.04, the value per cu. ft.
$0.04×27=$1.08, the value per cu. yd.

Small pressed steel pans of about one and one-half inches in diameter on top, and one inch in diameter on the bottom, and one-half inch deep, made in the same shape as the gold pans, are very convenient for holding and drying the values after they have been concentrated in the gold pans. Small porcelain crucibles are also used for this purpose. Small stoppered glass vials are sometimes used, especially if the samples are to be transported to a distant place to be assayed and weighed.

The box used for measuring the dirt to be panned can be of any convenient size, the cubical contents of which are known. A very convenient size is a box whose inside dimensions are exactly 6 inches long, 6 inches wide, and 6 inches deep. This box, when filled smooth to the top, holds exactly one-eighth of a cubic foot, or one-two hundred and sixteenth part of a cubic yard.

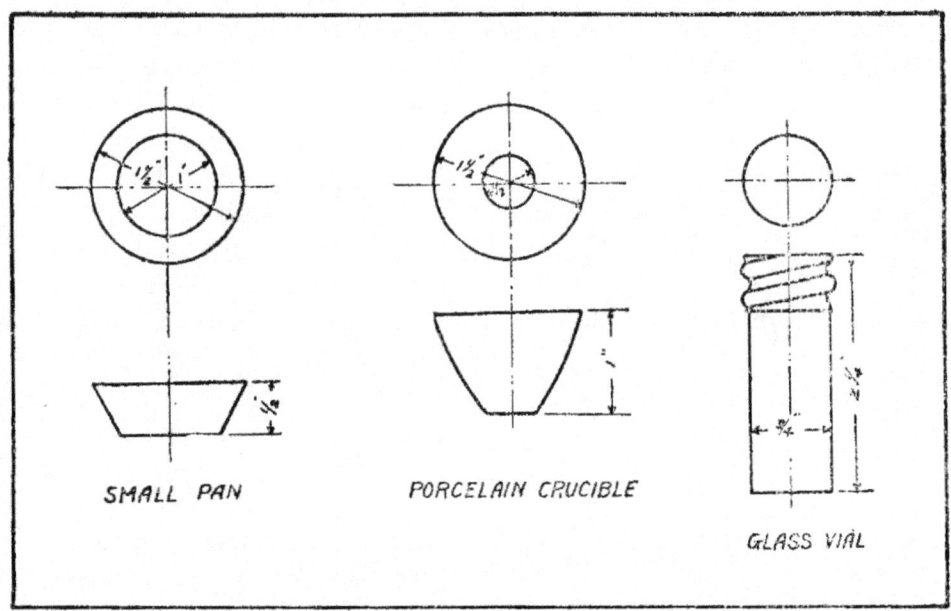

SMALL PAN   PORCELAIN CRUCIBLE

GLASS VIAL

A good set of balances with a standarized set of weights should be used for the weighing of the values recovered. It would be poor business to do all of the work well and then use an inaccurate set of balances to do this very important part of the work.

For identification purposes a tag must accompany each sample, and the concentrate from it. This tag always bears the number of the original sample from which it was taken and all additional data which applies to this part of the sample only, such as the number of boxes of dirt washed to obtain the concentrate, the method used in the testing, to obtain the sample, the condition and kind of the dirt, etc., etc.

WEIGHING OF THE SAMPLE.

After the sample is brought to the place where the testing is to be done, the measuring box is filled and weighed. The weight of the sample is the total weight minus the weight of the empty box or receptacle. Since the cubical content of the box is known, the weight of a cubic yard of the dirt can be calculated. This is needed in order to make the calculations from the gold per ton to the gold per cubic yard.

FIRE ASSAY OF PART OF THE SAMPLE.

At this point in the work a part of the original sample should be taken to be assayed by the fire method. This will give the exact mineral content of the sample, and from the value obtained the mineral present in the deposit can be calculated. The assay values are only used for calculations for the gold present, and are always given by assayers in ounces Troy—per ton (2000 pounds avoirdupois) of dirt, so that from this it is necessary to calculate the value per cubic yard of dirt.

PANNING.

One or more boxes of the dirt, as desired, are then dumped into a gold pan and panned, the amount of dirt from each sample which is panned to get each concentrate being recorded on its tag and in the sample book. This operation needs no explanation as those connected with placer mining are usually expert at this class of work. After all of the dirt is washed out and only the concentrates remain, there are two common methods of procedure for separating and collecting the gold from the black sand and other impurities which are always found in placers. One of the methods is to assay the concentrates by the fire method and weigh the resulting gold bead. This gives the amount of gold recovered from the amount of dirt panned, and the gold recoverable from a cubic yard can be easily calculated. This method is perhaps the quickest, most accurate, and cheapest of any. The other method is to add mercury or quicksilver to the concentrates and amalgamate the gold. The amalgam is then retorted. This operation removes the quicksilver and leaves the gold in the retort, from which it can be removed and weighed.

The values in this operation are for the gold recoverable from a cubic yard of the dirt, and are only used in the calculations for the gold recoverable. The practice of washing the black sand and other impurities in a horn is not recommended, as much of the value may be washed away and lost, thus giving an unfair result.

TESTING WITH THE DRY WASHER.

The same general method of procedure can be followed as above outlined, if the sample is to be tested by the dry washer in the actual testing. A sample of at least 100 pounds is needed.

From the above operations the value of the gold present and recoverable per cubic yard is determined for each sample. It is then necessary to determine the gold present and recoverable from the entire deposit. This work requires the services of a man experienced in engineering calculations.

CALCULATIONS FOR THE VALUES OF THE DEPOSIT.

Two separate and distinct calculations should be made to determine these. One is for the gold actually present, and for this calculation the value as determined from the fire assay of the original sample before it was panned is used. The other calculation is for the gold recoverable, and in this calculation the values obtained after the samples have been panned are used.

GOLD PRESENT.

The gold present in a deposit is found in the following manner: The cubical volume of each section is calculated from the cross sections which are taken from the topographical map. The depths used are as found in the test pits, except they should be adjusted as determined by the contours of the bedrock.

The gold present for a section is the product of the cubical contents of the section multiplied by the gold value per cubic yard, as determined by the fire assay of the original sample taken from the test pit for that section before the sample was panned. The gold present in the entire deposit is the sum of gold present in all the sections.

GOLD RECOVERABLE.

The same method as described above is used for determining the cubical contents of each section. The gold recoverable for a section is the product of the cubical contents of the section multiplied by the gold value per cubic yard, as determined from teh results obtained from the panning or dry washer tests.

After the above results have been obtained it is usually desirable to make a valuation of the deposit.

VALUATION.

The purpose of the valuation of a placer deposit is to find out the net profit obtainable after all charges have been deducted.

As there are no two placer deposits exactly alike, and as all differ so much from each other, it is impossible to give anything but a few suggestions as to what has to be taken into consideration in this matter. In addition to the estimation of the gold recoverable by the process of extraction decided upon, management, climate, labor, power, fuel, water, transportation, equipment, food supplies, interest on investment, depreciation, government regulation, kind of government, etc., etc., are some of the many items which are important and must be taken into consideration in the calculations for making the valuation of any placer deposit. For the above reasons it seems advisable to have a thoroughly competent, reliable and experienced engineer supervise and be made responsible for the valuation of a placer deposit. The work will not only be correctly done, but the report on the deposit as submitted by the engineer will be accepted by everybody as authoritative.

University of Arizona Bulletin

SAMPLING SERIES NO. 2   SEPTEMBER 20, 1917

THE SAMPLING AND ESTIMATION OF THE METAL PRESENT IN A DUMP OR TAILINGS HEAP

BY GEORGE R. FANSETT

The sampling and estimation of the metal present in an ore dump or a tailings heap are matters which are of considerable importance to most men connected with the mining industry, as they are often called upon to handle this class of work, either for themselves or for others.

There are several important points which can be settled concerning the material in an ore dump or tailings heap before any work is started in the shipping or in the treatment of the material, and most of these points can be decided from the results obtained from the analyses, the testing of the samples or the reports from the smelters on the samples.

A few of the important points which can be decided are the following:

1. The amount and kinds of valuable minerals present and their values.

2. The net profit, if any, which can be derived from the dump. This is the amount of money left over after all charges which may be levied against the material in the dump have been deducted. Some of these charges are the following: Treatment charges, freight, labor, tools, etc. As the profit to be derived is the most important matter concerning every dump, every possible precaution to include every charge which might be levied against the material should be taken into consideration in the calculations which will settle this point.

3. The process of treatment. The results from the samples should decide if the material making up the dump should be treated or dressed first, or if it is better to have it smelted as it exists in the dump.

Under the above headings are many subdivisions, which enter into the calculations, a few of which follow:

(a.) 1. *For Tailings.* The best ore dressing process of treatment for the extraction of the values from the tailings. 2. The kind and cost of the necessary equipment.

(b.) *For ores where it is desirable to concentrate the values before smelting*. A great part of the mine-run falls in this class. Among these are ores which will not pay the charges levied against them unless their values are concentrated before they are shipped or smelted. There are others where the net profit is much increased by this preliminary treatment.

(c.) *For ores which are high grade enough to smelt as they exist.* Under this heading several important points can be settled. (1) Whether it is better to smelt the ore locally, or (2) to ship it to a custom smelter, and if it is better to ship tp a smelter, (3) which smelter will give the best financial returns.

The above points are mentioned only to illustrate the nature of a few of the important matters that can be settled from the results obtained from the samples taken from an ore dump or a tailings heap, and will serve to emphasize the importance of the proper sampling of an ore dump or a tailings heap, and the points that can be settled from the results obtained in the analyzing and testing of the samples before the work is started on the shipping or on the treatment of the material making up the dump or heap.

It is far better to spend a few dollars in the sampling and testing of the samples than to ship the material to a smelter and discover that the money received will not pay the freight and other charges levied against it, or for one to install expensive machinery and find later that the expected values are not present or that the process of extraction is not adapted to that particular ore, or that a different process would have been better. In other words, it is better to be *sure* than sorry.

For reasons such as these, it would seem to be to the advantage of all those interested in the mining industry to understand the methods generally used in the sampling and in the estimation of the metals present in ore dumps and tailings heaps so that they will know how to do the work if called upon to do it. The purpose of this Bulletin is to indicate and describe the methods which are used by many of the large mining companies and many engineers for getting this preliminary information and doing this kind of work, in cases where a large expenditure of cash and expensive machinery is out of the question, and where the value of the dump is to be estimated within reasonable limits.

Before describing the methods of proceedure for doing the work, it may be advantageous to define several terms which are used in this Bulletin

Definition of an Ore.—*Richards.* An ore is a natural aggregation of minerals from which a metal or metallic compound can be recovered with profit on a large scale. When the per cent of metal is too low for profitable extraction, the rock ceases to be an ore. The rock has to be tested to determine this point.

Definition of an ore dump. An ore dump is a pile or heap of ore. The ore making up a dump is usually selected roughly for each particular dump,—that is, high grade ore is usually dumped in one pile, medium in another, and waste discarded.

Definition of a tailings heap. A tailings heap is a dump which is made up of the detritus or rejected crushed material from a metal extraction or a reduction plant.

Definition of a sample and sampling. A sample is a collection of fragments or pieces from a deposit which contains exactly the same minerals in exactly the same proportions as they exist in the deposit from which the sample was taken. In this bulletin the material saved from each cutting down operation is referred to as the sample. The act of collecting these pieces is called sampling.

Definition of the minerals present. The minerals present are those actually existing in the sample and the amount of each present is determined by a quantitative analysis of the sample.

Definition of the metal recoverable. The metal recoverable is that which can be actually recovered from the ore by the use of the processes of ore dressing or reduction utilized, or both, and as there are always losses in these processes, the metal recoverable is always less than the metal actually present in the deposit.

Definition of "values." Gold, silver or other valuable minerals which are present.

Testing. This term as used in this article may mean assaying, analyzing, tests for a process or, in fact, any or all of the various tests for which a sample may be used.

CAREFULNESS.

Sampling is slow, hard, caretaking work, and the greatest care is absolutely necessary and must be taken in every part and detail of the work . It would be poor business to spend a large sum of money and labor in the taking of samples and then find out that, owing to some part of the work being carelessly and incorrectly performed, the samples are worthless. In a case like this the results may be considered dangerous, as they may lead to heavy unmerited expenditures, and much money would be wasted. In order that the estimate of an ore deposit shall be correct, the figures which are used in making the estimate certainly must be correct. For the above reasons, careless, slipshod work is absolutely out of the question in taking samples and sampling.

RECORDS.

In the sampling of ore dumps and tailings heaps, it is necessary to keep a good set of records so that if any questions arise, at the time or after the work has been finished, they can be immediately settled and answered correctly and definitely. For doing this, it is well to use a topographical map of the deposit, a sample book and a diary. All matters and information pertaining to the deposit should be recorded in one or all of the above mentioned records. If any point to be recorded seems to belong to more than one of the above mentioned records, record it in all the records where it seems to belong. If it seems to belong to all, record it in all, and do not consider it useless repetition, as any matter worth recording is worth finding easily and quickly. These records represent the work done, and for this reason should be most carefully and accurately kept.

THE TOPOGRAPHICAL MAP.

As soon as a thorough preliminary examination of the dump or heap has been completed, a topographical map is made of it. It should be made to a scale large enough so that any and all details can be plainly marked on it. All distances for this map are measured on the horizontal; all elevations are calculated from a known and permanently fixed bench mark or datum. The map is used for locating the points where samples are to be taken, furnishing the data for making the calculations for the cubical yardage of each section of the deposit and for the whole deposit. When a test pit or a crosscut or other work is decided upon, and work started on it, the location of the pit or excavation, together with the number given to it, should be marked on this map at its proper location. Also when a test pit reaches the bottom of the dump, the elevation of the bottom is marked at the proper location. The depth of the dump at that point can then be found by subtracting the elevation of the bottom from the elevation of the top at that place.

On this map is also recorded by the numbers given to them the location of each of the samples. After these samples have been assayed, the values obtained are also marked. In fact, this map is used to record everything of this nature concerning the deposit.

The Sample Book.

Supplementary to the topographical map and diary, a sample book should be kept. In this book all details and records pertaining specifically to each sample are recorded. The following will illustrate some of the matters which are taken care of in the sample book: (1) Number given to the sample; (2) Date when the sample is taken; (3) Name of the deposit from which the sample is taken; (4) The location at which the sample was taken; (5) Assay returns from the sample, etc., etc.

A convenient form of sample book is one which has a page which is perforated near the far end so that this end part of the page or tag can be easily torn off and put into the sack with the sample to identify it. The number which is given to the sample is all that is usually written on the tag. This number, together with all other matters concerning the sample, is written on the part of the page which is fixed in the book and these pages form a complete record of the samples.

The form on Page 6 represents one of the pages used in a sample book, but this form should be changed, if necessary, to meet the requirements of any particular job.

The Diary.

Together with the aforementioned records, a diary should be kept. All other matters of importance which are not naturally included among those recorded on the topographical map or in the sample book are taken care of in the diary. Such matters as when the work is started on a certain pit, when and to what extent a pit caved, etc., etc., should be taken care of in the diary. As before stated, any inquiries which may come up in regard to the deposit, the sampling, or the samples from a deposit should find a satisfactory answer either in the diary, the sample book or on the topographical map, or all of them.

Numbering of the Excavations.

Each excavation is given a different number and in most cases they are numbered consecutively. This serves to make it easier to remember where each particular excavation lies. The one important point is that no two excavations have the same number, thus avoiding any confusion.

THIS PART REMAINS IN SAMPLE BOOK   TAG

THIS TAG IS PUT IN SACK WITH SAMPLE

_____ 19 _____ SAMPLE NO._____

NAME OF DEPOSIT _____

HOLE NO._____

KIND & CONDITION OF DIRT _____

WEIGHT OF DIRT PER CU. YD. _____

ASSAYS   GOLD.   _____ METALS

OUNCES PER CU YD.	VALUE PER CU YD.			

SAMPLE NO. _____

REMARKS

_____

_____

BLANK PAGE FOR A SAMPLE BOOK

Numbering of the Samples.

Each sample is given a different number, irrespective of its location, and by this number the sample is identified at all times with the aid of the notes kept in the sample book. No two samples should have the same number or mark. If a sample is concentrated, the concentrate or tailing should be given either an entirely different number, and notes made in the sample book as for any other sample, or if the same number is given to it, a note should be put on the tag stating just what it is. For example, "Wilfley concentrate, from sample No. 276."

When the sample is taken, the number given to it is marked on the topographical map at the proper location. This number is also marked on a page in the sample book and on the detachable part of the page or tag of the sample book. After the sample has been cut down to the desired size, the sample, together with the tag bearing its number, is put into a sack or other container.

Tags.

Only one tag is put into each sample sack with each sample, and the number on the tag, together with the notes kept in the sample book, serve to identify the sample at all times. The tags used for this work are made either of paper, soft wood or metal. Paper tags are commonly used except with samples which are very wet. These tags are usually the detachable part of the page of the sample book referred to, but can also be any piece of paper with the proper number written on it. These paper tags are rolled up tightly in the form of a lead pencil, and have their ends well crimped. This is done so that they will not unroll easily and get soiled, thus keeping the writing legible.

Metallic tags, having the number stamped thereon, are sometimes put into the sample sack with the sample instead of the paper tag. These are very serviceable with samples that are wet.

Soft wooden tags with the number written thereon with a hard lead pencil are also serviceable, especially with wet samples. A hard pencil is recommended for writing the numbers on the wooden tags, because the lead will cut into the wood and the indentation will remain even if the lead is rubbed or washed off.

Sacks and Containers.

Only new sample sacks should be used. If tin or other containers are used, they should be thoroughly cleaned out before the sample is put into them. Sacks which have been previously used for holding samples or dirty containers may contain values from the former samples which they have held, and these values will get mixed with the sample and enrich or salt it, thus spoiling the sample. It would be poor business to spend a large sum of money on a collection of fragments from a deposit in the collecting or taking of it, testing, freight, and other charges, if the collection is not a true sample, having been spoiled by the enrichment or salting from an old sack, when new sacks cost but a few cents each.

SIZE OF SAMPLE.

The final size of the sample depends on what is to be done with it. If it is only to be assayed or analyzed, only a few pounds are needed. The size of the sample sent to a commercial assayer or chemist for the common analyses does not have to be more than a pound. If the sample is to have a complete analysis made of it, a few pounds is usually sufficient. If it is to be tested for an ore dressing process and a process of reduction, from 500 pounds to several tons may be needed. In other words, the needs will determine the size of the sample saved. When sending a sample for analysis, it is well to keep a part of the sample sent, so that check analyses can be run on it if it is desired.

LIMITS OF SAMPLING.

Most dumps which are to be sampled will not stand a big outlay of cash for doing this work. Expensive machinery and power are usually out of the question. In many cases shovels are the only tools available. For this reason it is the practice of engineers to use methods for doing the work that will give approximate results in the shortest time, and in the cheapest manner possible. The estimates which are made from the results thus obtained should be correct within reasonable limits of each particular case.

The work which has to be done on most dumps before the shipping or treatment of the material should be started, to obtain this preliminary information, can be divided into several parts, the most important of which are the following:

(1) Taking the sample.
(2) Testing of the sample.
(3) The estimation of the values present and recoverable.
(4) The valuation of the dump.

These matters will be handled in this bulletin in the above order.

APPLICATION OF METHODS USED.

The methods of procedure in the sampling of ore dumps is practically the same as that used in tailing heaps, and unless some part of the work is mentioned as applying particularly to one or the other of these classes of dumps, the methods described will be understood to apply to the two classes of dumps.

In cases where it is questionable whether the dump is of value or not, the usual course of proceedure is first, to take grab or pipe samples. These are assayed and the results from them are used only to indicate whether or not a more thorough sampling of the dump is merited. To take grab samples, the dump is laid off in squares. Handfuls or shovelfuls of the material are taken, as fairly as possible, at each intersection of the lines forming the squares. This is usually all mixed together and assayed. The results from the assay are rough, and indicate only whether or not the dump is worth bothering with at that particular time.

In tailings heaps, pipe samples are sometimes used for this purpose instead of grab samples. This method can only be used in finely crushed material, and is done by driving a short length of pipe (1½-inch pipe answers) into the heap at points from which the samples are wanted. The pipe, with the sample in it, is withdrawn and the sample is knocked out of the pipe and assayed. These samples, as in the case of the grab samples, are rough. If the results from these rough samples indicate that it is worth while more thoroughly to sample the dump, there are many methods of proceedure for so doing, among which the following are often used:

Ore dumps: from 500 lbs. to about 5 tons. Fraction sampling or crosscuts. Above 5 tons—crosscuts, test pits, or drill hole sampling. Tailings heaps: Above 5 tons—test pits or drill hole sampling.

METHODS OF PROCEEDURE.

A topographical map of the dump should first be made. From this map the cubical contents in the dump can be calculated. The tonnage or weight of the material in tons can be found by multiplying the weight in tons per cubic yard by the number of cubic yards. The weight can be found by weighing a known quantity as a cubic foot.

Small dumps, or those containing not over 5 tons, can be sampled either by using what is commonly known as the fractional method or by crosscuts.

The fractional method is one commonly used when the ore is to be shipped to a smelter or to a reduction plant. By this method all the material in the dump is shoveled from where it lies to a different place. Every second, third, fifth or tenth, or any other numbered shovelful decided upon in advance, is shoveled into a separate heap on a clean, tight platfform or other clean, smooth surface, and the heap thus made up is kept for the sample. The most important points to be taken care of are: (1) If every fourth shovelful is to be saved for the sample, be sure and save only every fourth and no others, or if it is decided to save every tenth, be sure and save only every tenth shovelful of the material. (2) Do not pick the material which is to be saved for the sample. The shovelful saved for the sample should contain as near as possible the same amount of material and the same kind of material as that previous and following. This work separates the dump into two lots: one is to be used for the sample and the other is to await the results obtained from the testing of the sample.

Where the material in the dump does not cave, sampling is sometimes done by the crosscutting method. The width of all of the crosscuts must be the same throughout their entire lengths, and from top to bottom. The material taken from the crosscuts is all mixed together and is used for the sample. This method is rough, but in cases where a quick, cheap and fairly accurate estimate is desired, it can be used to advantage. The sketch on Page 11 illustrates this method.

Crosscutting is usually impractical in tailings heaps, owing to the fact that the material usually caves easily, and if the crosscuts are broader at the top than at the bottom, there is more material taken for the sample from the top, and for this reason the sample is sure to be unfair unless all of the material in the dump is absolutely uniform. Tailings heaps not larger than 5 tons are seldom of any commercial value.

For a small dump the only tests usually desired are the assays and smelter reports. For these tests 20 pounds is usually more than is needed. When the amount saved for the sample from the first operation is more than 800 pounds, and the amount desired for the final sample is about 20 pounds, one of the following methods can be used in cutting down the sample to the desired size.

METHODS OF CUTTING DOWN SAMPLE FROM ORE DUMPS WHEN QUARTERING MACHINERY IS NOT AVAILABLE.

Down to about 500 pounds. Fractional method of cutting.
From about 500 pounds to 100 pounds. Coning, quartering by use of a cross or the Jones sampler.

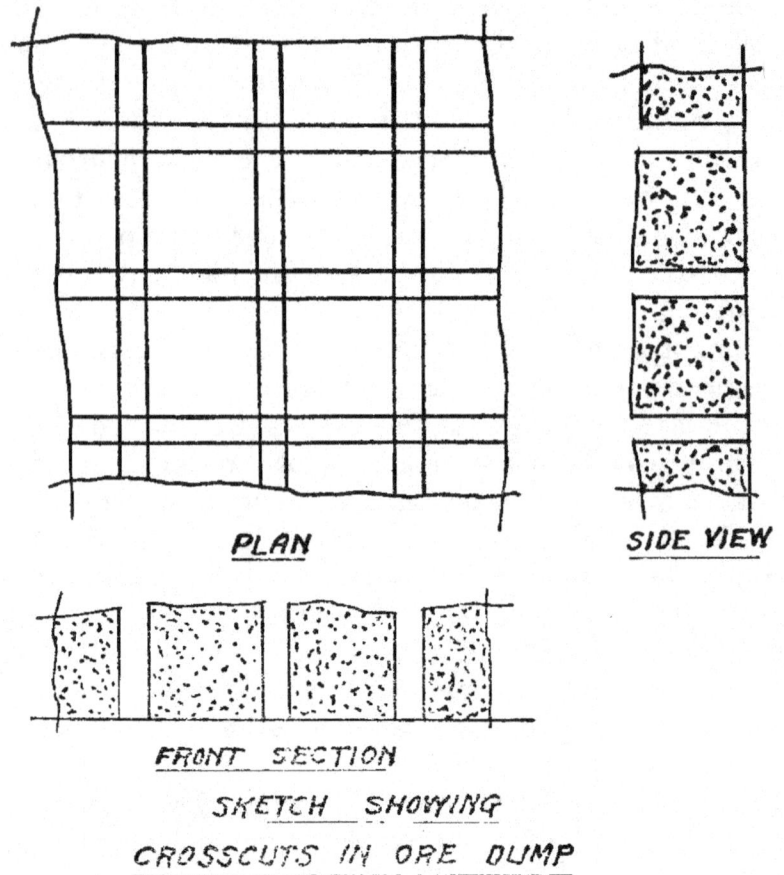

PLAN SIDE VIEW

FRONT SECTION

SKETCH SHOWING

CROSSCUTS IN ORE DUMP

From 100 pounds to the size desired. By use of canvas or table oilcloth, or the Jones sampler.

TAILINGS HEAP SAMPLE.

Down to 500 pounds. Fractional method of cutting down or by 'use of the Jones sampler.

From 500 to about 100 pounds. Coning; quartering by the use of a cross, or the Jones sampler.

From about 100 pounds to the size desired. Method using canvas or table oilcloth, or the Jones sampler.

The reject from each cutting down operation is usually added to the pile left from the first cutting down operation.

There are many cases where no machinery is available and shovels are the only tools to be had for cutting down samples to the desired size. In such cases it is necessary to use a method for cutting down the sample which will give results which will be, within reasonable limits, as accurate as possible. In these it is a common practice to use the fractional method repeatedly, as was described under "The Fractional Method of Sampling," for cutting the sample down to about 500 pounds. From 500 pounds to about 100 pounds the coning method is often used.

When the coning method is used, the pieces of material making up the sample must not be larger than will pass through a 2-inch screen when the sample weighs over 300 pounds. When the weight is less than 300 pounds, all of the pieces in it must pass through a 1 inch screen. It is therefore necessary to crush or break up all pieces which are larger than indicated above. A cobbing hammer and anvil are convenient for breaking these pieces up if a crusher is not available. If a crusher is to be had, it is better to crush all pieces to less than one inch.

The main advantages of the coning method are: (1) No expensive machinery is needed; (2) Any kind of mineral can be cut down by this method.

The main disadvantages of this method are that the material has to be handled so many times that the cost of the work is very high, and it is next to impossible to get an absolutely even distribution of the values.

The last heap of material saved for the sample from the preceding cutting down work is leveled to a circular form by use of a hoe, flat-nosed shovel or similar tool, so that it is not over four inches deep. The next step is to cone the material ,which is done in the following manner. From the outside part of the leveled heap of material, at points equally distant from each other, equal amounts are shoveled up and allowed to fall onto the center of the leveled heap in such a manner that the material is evenly distributed on all sides of the cone which is formed. In this way, only a portion of the heap is shoveled up in passing once around the heap, the metal is mixed, and a fair distribution of the values is accomplished. When all of the material outside of the cone, which is formed at the center of the heap, has been shoveled up to the cone, all of the fine material left outside of the cone on the platform is swept into a shovel and shoveled onto the top of the heap. The cone is then shoveled to a different platform, or to another part of the same platform which has been thoroughly cleaned. It is then leveled again to a circular form and coned again as was described above. This process is repeated until the material has been thoroughly mixed. When this has been accomplished, the last cone formed is leveled again to a circular form, the depth of the material being about four inches.

The leveling of the final cone to a circular form should be done very carefully, as usually the finest particles lie close to the apex of the cone, and as these usually carry the highest values, they should be distributed as evenly as possible in each of the four quarters.

The leveling is ordinarily done by the use of a shovel, the back part of which is held vertically toward the apex of the cone. It is the common practice to start at about one-half of the distance from the outside of the cone and its apex, and work around, always working toward the apex, the material being dragged out over the outside edge of the cone. After the material is leveled evenly, it is divided into equal quarters by cutting along two diameters which are at right angles to each other. The two opposite quarters are kept for the sample; the other two are discarded and are added to the balance of the material which made up the original dump.

The sketches on Page 14 illustrate the above.

One operation of this method thus cuts the sample down one-half. If the sample is still too large (over 100 pounds) this entire operation is repeated on the portion saved for the sample until it reaches about 100 pounds, when the method using table oilcloth or canvas is more convenient. Much care must be taken in doing this work, as well as to clean thoroughly the parts of the platform where the discarded material has been, in order to prevent salting from the values which may have been left there. Table oilcloth is considered one of the best materials for this purpose, as it has a perfectly smooth surface, and for this reason no values can get into the fibres or cracks and thus detract from one sample or salt another.

When this method is used, all pieces of the sample must be small enough to pass through a one inch screen. The final sample saved from the former cutting work is shoveled onto a piece of the oilcloth about six feet square. Then taking the two opposite corners of the cloth, one in each hand, one corner is lowered at the same time and the same rate that the other is raised, the bottom of the cloth always resting on the platform. This motion rolls and mixes the sample. When this has been carefully done, the two opposite corners are taken in the same way, the sample is again rolled and mixed, but in the opposite direction to the first operation. This is all repeated several times, until the sample has been thoroughly mixed. The canvas is then spread out flat on the platform and the sample leveled to a circular form not over three inches deep, as was described before in connection with the coning method. It is then divided into equal quarters

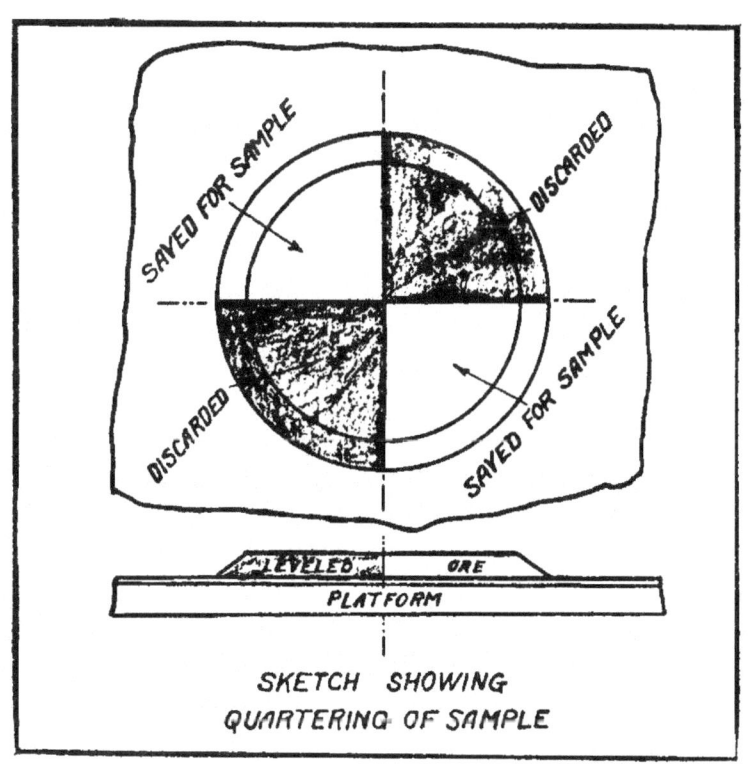

SKETCH SHOWING
QUARTERING OF SAMPLE

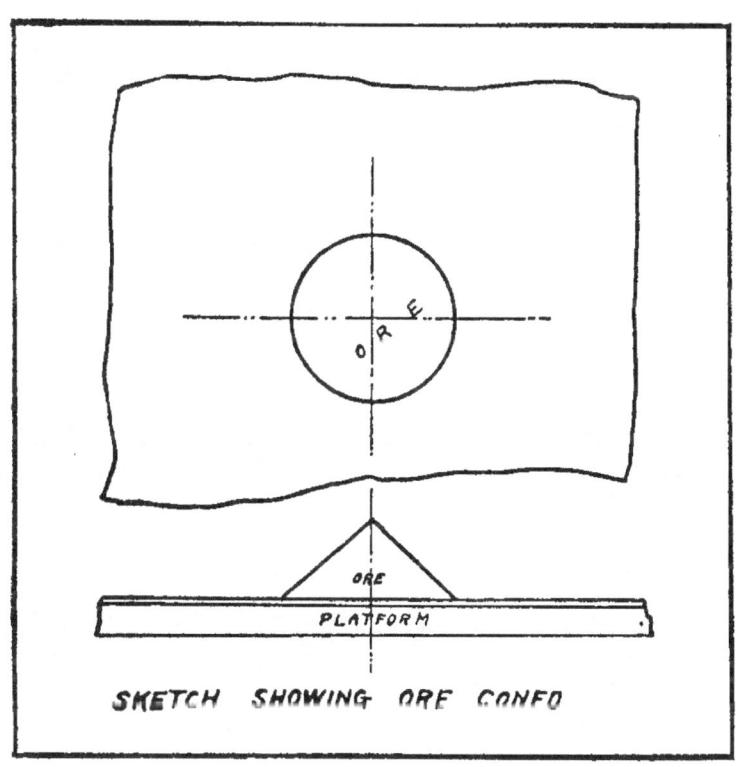

SKETCH SHOWING ORE CONED

by cutting it along two diameters at right angles to each other, as illustrated above in Sketch No. 3.

The opposite quarters are kept for the sample and the other two discarded. The place where the discarded material has been should be thoroughly cleaned, so that none of it will be added to the sample. The final sample is then thoroughly mixed, put into a sack with the tag bearing its number and is ready to be analyzed.

There are several other methods than those explained above which can be used to advantage in all or in parts of the work of cutting the sample down to the desired size. Some of these methods are considered by many engineers to give more accurate results, and in some cases the work can be done more rapidly and cheaply. The main drawback is that special apparatus is required, but when this can be procured they will give more accurate results.

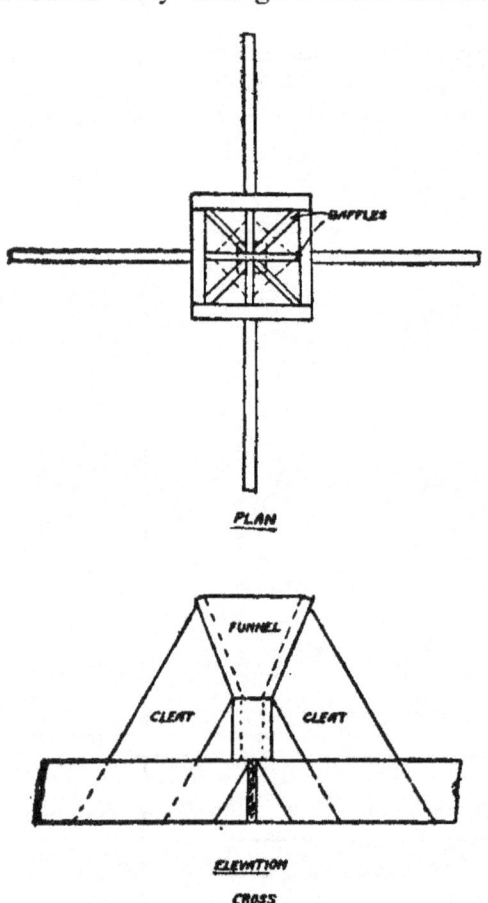

PLAN

ELEVATION

CROSS

FOR

QUARTERING DOWN SAMPLES

One of these methods is the use of a cross for quartering the material. This is used after the sample has been cut down to about 800 pounds by the fractional method. When the sample weighs over 300 pounds, the pieces should be small enough to pass through a 2-inch screen; when it weighs less than 300 pounds, they should pass through a 1 inch screen.

The cutting down by this method is done with the aid of an apparatus, a sketch of which is shown here. It consists of four arms which are built at right angles to each other in the form of a cross, with a funnel located above the cross, the center of the funnel and the spout from it being exactly over the intersection of the arms. The spout should be long enough and small enough so that the mineral will fall vertically, or straight down, thus accomplishing an even distribution.

This apparatus is placed on a clean, tight, level platform; the sample saved from fractional sampling, after being thoroughly mixed by coning, as explained above, is then shoveled into the hopper of the funnel, care being taken that none of the sample falls over the side of the funnel into any one of the quarters. It is a good precaution to cover the arms of the apparatus to prevent this, and after each run has been completed, to collect the material which has fallen over the side of the funnel, and shovel it into the funnel, thus adding it to the sample to which it belongs.

The material of the sample between the arms of the cross in the two opposite quarters, is saved for the sample, as was explained under the coning method. That in the other two quarters is discarded. The space between the arms of the cross where the rejected parts have been, is well brushed and cleaned off, so that any values which may be there will not salt the next sample. One operation thus cuts the sample in half. After this is well done, the apparatus is lifted and placed at another clean part of the platform and is ready to have the above operation repeated if the sample is still too large.

When the sample reaches about 100 pounds, it is well to use the oilcloth method, or a Jones sampler, if such is available. At this stage there should be no pieces of the sample which will not pass through a one inch screen.

THE JONES SAMPLER.

The cutting down of a sample from any size can be accomplished rapidly and accurately by the use of a Jones sampler, providing the size of the pieces in the material is not more than three-quarters of the width of the slots in the sampler, a picture of which is shown below. This is an inexpensive apparatus which stands rough handling, and gives good results providing it is properly used.

On top, the sampler has a row of horizontal slots, all of which have the same length and width. From each one of the slots runs a chute, every second chute running in the same direction; that is, the outlet from the first chute is opposite that of the second, and so on. The material is shoveled slowly from a flat-nosed shovel or scoop transversely onto these slots, much care being taken that the chutes do not clog, and to distribute the material evenly by moving the shovel back and forth from one side to the other of the apparatus. The end of the shovel should be held about one inch above the slots of the sampler, and in this way equal amounts of the sample will fall into each slot and run down and out of the chute under it. A

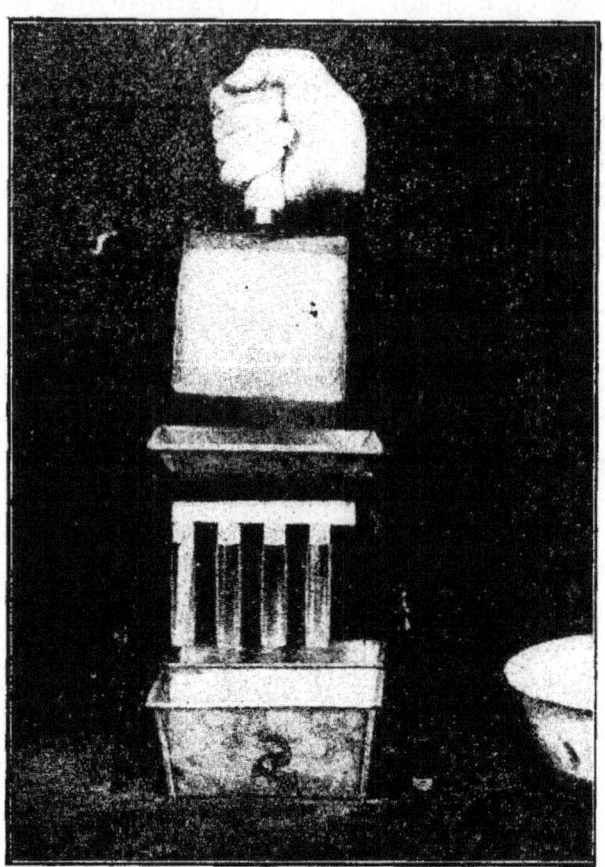

pan is placed under each row of chutes, to catch the material and if the work is properly done, one-half of the sample will be caught in each of the pans. That in one pan is saved for the sample, and the other discarded. The operation is repeated until the desired size is obtained. It is then sacked with the tag bearing its number, and is ready to be tested.

In some cases it is advantageous to bank up several samplers, one above the other, so that the sample from the first falls into the second automatically, and in this way the work is done much more rapidly and with less handling. When used in this way, care must be taken to have the samplers arranged in such a way that the material entering one sampler from the one above it is evenly distributed.

The Jones sampler is particularly adaptable for cutting down samples from tailing heaps, as these are usually crushed to sands. When the material has to be crushed, it is usually advisable, for the sake of economy, to use the fractional method to cut it down to about 500 pounds. The large pieces are then broken up and thoroughly mixed with the rest before it is put through the sampler. This can be cut down by using a large sized sampler, one having slots 1 to 1½ inches wide. When it has been cut down to about 100 pounds, it is advisable to use one with slots not over ½ inch in width. It is therefore necessary to break up all of the pieces larger than this size, so that they will pass through the slots. They should be thoroughly mixed either by coning or by use of the oilcloth or canvas, and shoveled into the sampler.

The sampling of ore dumps from 5 tons to about 100 tons can be done cheaply and conveniently by the cross-cut method described above. When this method is used, great care and judgment are needed in locating the crosscuts, so that as fair a sample as possible will be taken from the dump. The material taken from the crosscuts is the sample, and it can be cut down to the desired size by using the methods which were described above.

Dumps of this size can also be sampled by using test pits. These are pits which are sunk in the dump, the material taken from

them being the sample, which can be cut down to the desired size by methods previously described. The location of these pits is a most important matter, requiring much care and judgment so that as fair a sample as possible will be taken. This method is described very fully in the Arizona State Bureau of Mines Bulletin No. 51.

Another method often used where the material is not coarse is that of the augur drill. This method is also described in Bulletin No. 51 of the Arizona State Bureau of Mines.

Churn drills can be used if necessary, but as this is very uncommon practice, reference only is made to it here.

Sampling Dumps of Over 100 Tons.

The sampling of dumps of over 100 tons is usually done either by using test pits, augur drill or churn drills, as described above.

From the above it is evident that more or less judgment is needed in the selection of the process best suited for the sampling of any particular dump. In some cases a combination of the above methods is cheaper and better, but these points should be decided before the work of sampling begins.

ANALYSIS OF THE SAMPLE.

The next step after the samples have been cut down to the desired size is to have it assayed or analyzed. This part of the work should be done by a thoroughly competent and reliable man, since the results from these analyses will decide many important matters regarding the dump, a few of which are the following:

1. The amounts of the valuable minerals present per ton of the material in the dump. This is the weight and value of the gold and silver present, and the per cent of the copper, tin, zinc, lead, or any other minerals which may be present.

2. The amount of the various fluxes present. This is the per cent of lime, iron, silica, and other fluxes present.

The results from the analyses are used in the calculations which are made to determine an estimation of the minerals present and the valuation of the dump.

ESTIMATION OF THE TOTAL VALUABLE MINERALS IN THE DUMP.

This is the total weight of each valuable mineral which is in the dump, and is determined by multiplying the weight of each mineral present per ton of material by the number of tons of material in the dump.

The tonnage of the dump is calculated by multiplying the number of cubic yards in the dump by the weight in tons of the material per cubic yard. The number of cubic yards is calculated by using the cross sections of the dump which are taken from the topographical map of the dump. The weight of a cubic yard of material is found by using the weight of a known volume of the material as, for example, a cubic foot of it, and making the necessary calculations for a cubic yard. This part of the work should be done by a man who is familiar with engineering calculations as a wrong estimate would lead to serious consequences.

TESTING FOR A PROCESS OF EXTRACTION OR CONCENTRATION.

Where the assay returns and the estimate of the minerals present indicate that the dump is of commercial value at that particular time, in many cases, especially in very large dumps, it is desirable to have the material tested for an ore dressing process, or process of extraction. This is the process which is best suited for recovering the valuable minerals and rejecting the gangue.

It is the usual practice to send the sample to a competent and reliable man or company, who make a business of testing ore. These men know how to do the work and have the available machines and equipment for testing the sample thoroughly after the preliminary tests have been made.

VALUATION OF THE DUMP.

After the above points have been decided, the next matter of importance is the valuation of the dump. The principal point to be decided from this is the ultimate net profit which can be made from the dump. This is the amount of money which will be left over after all the charges which may be levied against the dump have been deducted. As this is the most important point to be decided, and will determine whether the dump is of commercial value or not, and to what extent it is of value, these calculations should be made by an experienced, capable and reliable man.

The following will serve to indicate a few of the matters which should be taken into consideration in making these calculations:

1. Amount of valuable minerals which are present, and the amount of money that will be paid for them.

2. Handling charges—Labor, freight, and all other charges of this nature.

3. Equipment—Tools, machinery and other necessary articles.

In addition to these, a few other items affecting the valuation calculations are management, climate, water, power, fuel, food supplies, interest on the capital invested, etc., any of which, if neglected, will give incorrect valuation.

If the valuation should determine that it would not pay to ship the material at that particular time, it is well to keep all the data and results obtained from the sampling, estimating and valuation, as they may be of use at some future date, when, for some unexpected reason, such as higher prices paid for the minerals, the discovery, perfecting or development of a process of extraction or other reason, it may be possible to handle profitably the material in the dump.

University of Arizona Bulletin

SAMPLING SERIES No. 3   OCTOBER 20, 1917

TAKING SAMPLES AND MEASURING THE WIDTH OF A MINERALIZED VEIN

By George R. Fansett

Taking samples and measuring the width of an ore body are matters which everyone connected with the mining industry should be able to do properly. The purpose of this bulletin is to describe the methods generally used by many engineers, so that one may understand how to do this very important class of work correctly.

The width or thickness of a vein at any point is the distance between the walls of the vein, measured along a perpendicular or at right angles to the plane of the vein at that particular point. For this reason there is only one width or thickness of a vein for each point or location in it. The following sketches illustrate this:

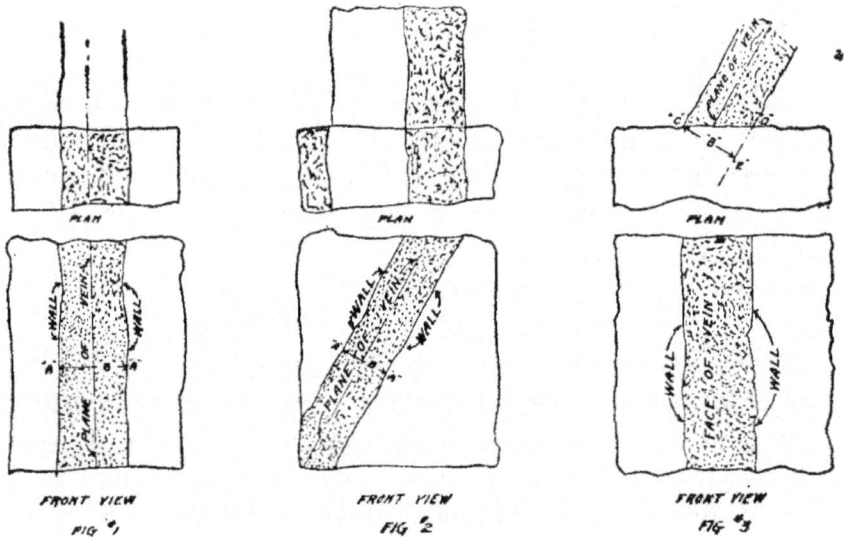

Figures 1 and 2 illustrate veins the faces of which are at right angles to the plane of the vein. In cases of this kind it is a simple matter to make this measurement as "B". In a case like that shown in Fig. 3, where the exposed face of the vein is not perpendicular to the plane of the vein, it is often difficult to make this measurement correctly. In such instances there usually arises a great difference of opinion as to the width of the vein, unless the measurement is properly made.

One of the best methods for making the measurement is to hold the Zero (°′) end of the tape or rule at the contact of the wall of the vein as at "C", shown in Fig. 3. With the tape or rule held perpendicular to the plane of the vein, the reading on the tape where the other wall, if projected as at "E", would hit the tape, will give the true width, "B"—("C" to "E") for the vein at that particular point. "C" to "D" is not the width of the vein.

In mining terms a sample is usually considered to be a collection of fragments or pieces from a deposit which contains exactly the same minerals in the same proportion as they exist in the deposit from which they were taken. The act of collecting these pieces is called sampling. The material from a cross section or core, from a vein at any point in the vein represents a true sample of the vein at that particular point.

Fig. 4 illustrates this.

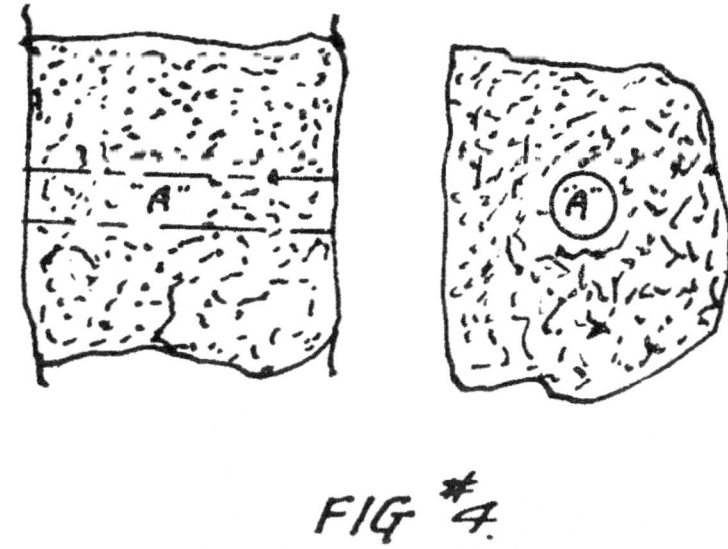

FIG #4.

All of the material or core taken from the hole "A" represents a true sample from the vein at "A", but at no other point in the vein.

The above represents ideal sampling, and can sometimes be approached by the use of a diamond drill or other similar apparatus, but owing to the high cost and other considerations, it is seldom possible to sample a mine in this way. For these reasons, most engineers use methods which are quicker and cheaper, with the intention of approaching in accuracy the ideal case as closely as possible.

For this work many engineers sample veins by chiseling or hewing a channel shaped groove across the width of the vein, catching the material which is hewn out and using this material for the sample for that particular point of the vein.

The following sketch illustrates how these grooves appear on the face of a vein after being cut.

Figure 5 represents a vein whose face is perpendicular to the plane of the vein. In either case the sample taken from the groove will be a fair sample, since the same proportion of minerals present is maintained. The only difference in the two cases is that the samples from No. 6 will be larger than from No. 5. The width of the vein should be measured, as explained before.

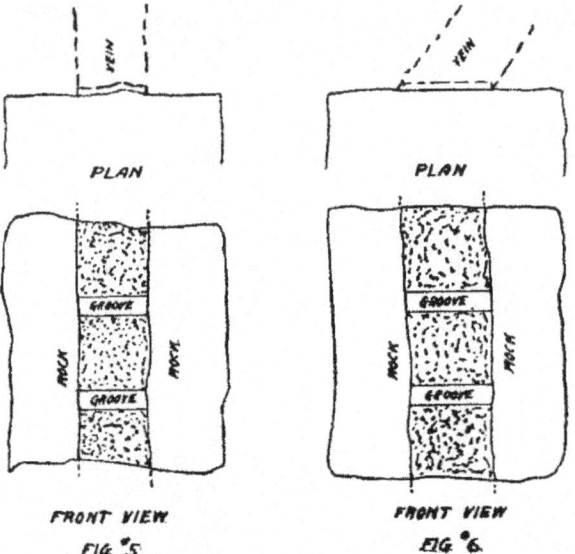

PLAN

PLAN

GROOVE
ROCK   ROCK
GROOVE

GROOVE
ROCK   ROCK
GROOVE

FRONT VIEW
FIG. 5

FRONT VIEW
FIG. 6

The groove which is chiseled out should have as nearly as possible the same width throughout its entire length. In most cases, a fair sample can be taken from a groove from three to six inches in width. The point must be decided for each particular sample, and will depend upon the hardness of the rock, the kind of ore deposit, the size of the sample wanted, and many other factors, each one of which will help to decide this important point.

The groove should have the same depth throughout its entire length if the face is fairly straight. In cases as illustrated in Fig 7, the groove should be chiseled to a greater depth nearing the center than at the sides. This is for the purpose of keeping the same proportion in the sample as they exist in the vein.

A fair sample can usually be obtained from a groove one-half an inch to three inches in depth. This, like the width of the groove can only be decided on the ground.

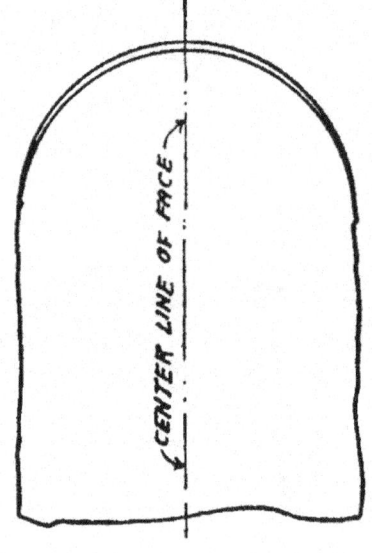

CENTER LINE OF FACE

FIG. 7

Engineers usually use moils or gads for hewing out these grooves. For striking the moil, a single jack (four pound miners' hammer) is very serviceable. If the vein matter is particularly hard, a double jack (eight to ten pounds, two handed, miners' hammer) is better.

The sample pick is seldom, if ever, used where great accuracy is wanted, and is not favorably looked upon by most engineers for doing this class of work. The reason for this is that it tends to pick out the softer spots in a vein. Since the soft material is usually the richer material of the vein, and the hard the leaner, it is easily seen that the sample take in this manner may be worthless as a true sample for that vein. The sample pick can sometimes be used to advantage by a thoroughly experienced sample man, when great accuracy is not desired.

The geologist's hammer is likewise not used for this work, as it tends to hit the projecting parts of the vein, and as these hard parts are usually made up of the leaner material, the sample may be worthless.

The practice of putting in a pop shot to break out the sample in no manner represents sampling. Even if the rock which is broken out by the shot is carefully quartered down, it cannot be considered a sample of the vein. The reason for this is that the shot usually tends to break out a conical shaped cavity whose axis is the drill hole and whose apex is the inside end of the drill hole. For this reason, a greater amount of the rock around the outer end of the drill hole will be broken than around the inside end. This, of course, would give a collection of fragments or pieces of rock that does not contain the same minerals in the same proportion as they exist in the vein, and for this reason, if for no other, it is valueless as a sample from the vein.

The practice of gouging out a specimen from the heart of a vein and giving the entire vein the values which have been derived from this specimen, of course needs no comment.

The grab sample for accurate sampling is likewise looked upon unfavorably by engineers, as one is almost bound to take the bright high valued pieces from a dump or deposit, if his eyes are open, and if he closes them he will usually get either too much of the fines or the lean rock. In any case, it is likely to be of no value for accurate sampling.

A powder box or other convenient receptical is useful in catching the pieces of the sample as they are hewn from the groove. It should be held as close as convenient below the cutting tool so as to catch all that is hewn out. Great care should be taken that none of the material making up the sample is lost.

For doing this work, at least two men are needed, one to do the cutting and the other to hold the box, keep the records and superintend the work generally. They should take the greatest care in every detail of the work, for, if one part is poorly done it may make valueless all of the work connected with that particular sample.

In some cases, especially when taking samples from an ore body whose ore leaches readily, or when the exposed face is very rough, it is desirable to break down and smooth off the exposed face. This makes it not only easier to get a true section from the vein, but in the case of minerals which leach easily, one gets back into the ore body proper, and is not so likely to get a high grade or salted sample.

In cases where the vein only occupies a part of the face of the drift, it is often better to take for the sample of the vein the material hewn out from the vein. The material from the hanging wall can be put in a different sack and that from the foot in another. It will thus be possible to analyze each separately if desired. This matter can be settled for each particular sample.

In cases where the vein is so wide that the sample will be altogether too large, it is convenient to measure off certain distances on the vein and take samples from them, records being kept so as to identify each sample.

The samples taken in this way are usually much larger than is needed for the assays or analysis which are to be run on them. In these instances it is necessary to cut them down. Methods for cutting down samples to the desired size, and for keeping the records of the work, were described in Arizona State Bureau of Mines Bulletin No. 63, Sampling Ore Dumps and Tailings Heaps.

MILL AND SMELTER METHODS OF SAMPLING
(By H. J. STANDER).

The idea seems prevalent among men operating small properties in the mining districts of Arizona that the smelters and custom mills do, in many cases, give false returns on ores that are sent in for smelting or treatment. With a view to ascertaining the facts of the case, the Arizona State Bureau of Mines has conducted an investigation and has found that it would be utterly impractical for any smelter or mill to so falsify its method of sampling as to give low returns and still give the shipper one half of the sample to allow check assays to be made.

This bulletin is written with the idea of showing the small shipper how to sample, so that he may sample his ores before shipment and ascertain the correct valu,e so that he may have a check on the smelter. Smelter methods of sampling are also given, so that the reader may determine for himself, that, owing to the great changes in the routine and the large number of men that would be necessary in any falsifying system, the smelters could not afford to give incorrect returns.

Moreover, it must be considered that when the Copper Queen mine, for instance, ships to the Copper Queen smelter, the smelter purchases the ore precisely the same as they do the ore of the man who ships ten tons, and the sampling and umpires are done under the same conditions.

It may or it may not be true that the smelters charge exhorbitant rates, or make undue penalties, but it is certainly true that the assays made for the basis of settlement are from correct sampling.

The question of the selling of ores and information relative to the necessity for smelting charges, penalties, etc., will be discussed in a later bulletin.

The sampling of the ore forms a very important part of the operation carried out in a mill or a smelter. The reasons for systematic sampling in a concentrating plant or mill are not perhaps always the same as those in the case of a smelter. Where a company concentrates and smelts its own ore, the sampling in the mill is carried out more for a check on the work and an indication as to what is being done than for any other reasons, although the sampling here is of just as much importance as in the case where the mill sells the concentrated ore to a smelter. In the latter case, both parties usually have their own sampling departments.

When a lot of ore comes to a smelter, it is highly essential to know its composition and value. The value is calculated from assays and analyses made on various samples of the given lot of ore. One can easily see that these assays and analyses should be as near the correct figures as possible, because a very light variation in such figures may alter considerably the total value of the ore lot. It is possible to assay and analyze quite accurately, and the results will be quite satisfactory, provided the samples used truly represent the lots of ore.

One of the chief problems, then, to be faced by the mill or smelter, is to obtain true samples of the ore. Such sampling must necessarily be accurate if ores or metallurgical products are sold by one and bought by another company, as the total value of the ore is always calculated from the assays and analyses of the samples. We have in the United States public sampling and ore-purchasing companies, who act as disinterested parties between buyer and seller, sampling the ore for a fixed charge per ton. It is very easy to see that these sampling companies greatly facilitate the selling and buying of an ore.

Most milling and smelting companies have their own sampling departments, and always sample an ore as it comes in, whether it had been sampled previous to its shipment or not. It is impossible for such a company to treat each ore separately, both because of the cost and because of technical difficulties. Mills and smelters work continuously, and a shut down, for however short a period, means a loss in labor, power, etc., and thus to allow one lot of ore to be heated altogether by itself is a practical impossibility. From this, it is quite clear that the only way of ascertaining the amount of the valuable metals in the ore lot is by assaying and analyzing true samples of the ore.

Before discussing in full the smelter methods of sampling, it may be well to note how sampling is usually carried out in an assay or metallurgical laboratory. As every lot or sample of ore comes into the laboratory, it has a number or name attached to it. The assayer keeps a very careful record of all data in a record note-book, in which he puts down the date received, name, number, and any other necessary information. Should the sample have no number, he gives it an arbitrary one, which can be identified in the future. After all this information has been recorded, he next finds the gross weight, and if the ore is wet, two samples are taken and the moisture determined. The ore is then thrown on the sampling floor. It is an easy matter at this stage to find the net weight of the ore, as it can be done by finding the weight of the sack, box or bottle in which the ore came, and subtracting this weight from the gross weight already found. At this stage some valuable information with regard to the character of the ore can be obtained, as the ore is still in a coarse condition.

Now comes the sampling, which is done by gradual crushing, mixing and sampling down. This sampling down is very often done by hand, and the method used is called the Cornish method. This method consists in coning the heap of ore on the sampling floor, then working it down into a circular shape and cutting it into quarters by making two diameters in the circle at right angles. The sample of ore has thus been thoroughly mixed and is divided into four parts, each part forming a segment of the circle. By removing two opposite segments, and then mixing the two remaining parts together again, it becomes possible to cone half of the original amount of the ore. After it has again been coned, the operation of quartering it is repeated, and the amount is again halved. This operation is kept up until a small enough sample is obtained. The Cornish method is a common way of hand sampling and since hand sampling is still somewhat used in the smelters, it is well to be aquainted with it.

The hand sampling, as carried out in the smelter is, however, usually of a different character. It is commonly done at the smelter when the ore is unloaded by hand. As the ore is unloaded from the cars, or transferred from one place to another, the workman throws, say, every tenth shovelful on the sample heap. By this method, one tenth of the ore goes to the sample heap, which is again sampled down.

A sample may be taken from a stream of ore while it passes from one place to another, and in such a case it is called a "running sample." Such a running sample is taken either by hand or by machine, but where the sample is not taken from such a stream of ore, the sampling is done by hand.

DIFFICULTIES IN SAMPLING

When an ore contains the metals in the native form, or in small quantities of very high-grade materials, carried in barren gangue, the task of sampling is quite a hard one. It is easy to see that wehn the metal values are finely disseminated throughout the mass of the ore, the sampling is not so difficult as when the metal occurs in large masses or crystals. When one considers that the amount of ore assayed is only very small, one can realize that, should this assay-sample contain a larger or smaller percentage of crystals of the metal than there actually are in the ore, the results would be very misleading. It is thus necessary that the ore shall be crushed fine enough to distribute the metal values as evenly through the ore as possible, for even after such an ore has been crushed to 70 mesh, it can easily happen that two samples, when assayed, will vary as high as 20 per cent. Such an ore frequently contains very rich particles of metal, which break off as crushing proceeds, and it is thus possible to come to a point where one has equally sized particles of nearly pure metal gangue. To get a true sample, it is thus necessary to continue the crushing until this point is reached.

REQUISITES TO ACCURACY

The first requisite to accuracy in sampling is that the sample and reject shall be uniform in composition at each stage of division. In order that this can be the case, there must be perfect mixing and very accurate dividing. After the sample has been cut down it should be recrushed sufficiently so that the ratio of the weight of the sample to the diameter of the largest particle in the sample shall not be below a safe proportion. If these conditions are carefully adhered to and thorough cleanliness is practiced throughout the entire operation, the limit of error is brought as low as possible.

The various ways of hand sampling will not be considered.

FRACTIONAL SELECTION BY SHOGEL

When the ore is moved by a shovel, either to load or unload it, every fifth, tenth or twentieth shovelful is thrown aside for a sample, as already noted above. How much of the ore shall be required for a sample—in other words, whether the workman shall throw on to the sample heap every fifth, tenth or twentieth shovelful, is always decided by the richness of the ore and the distribution of the minerals in it. In the same way, when loading or unloading ores in sacks, every fifth, tenth or twentieth sack may be set aside to form the sample. It is now necessary to crush this sample, and after it has been crushed the same operation of taking, say, every fifth shovelful, can be repeated. This alternate cutting down of the sample and crushing it is kept up until a small enough sample is obtained. The Cornish method of sampling, as used in an assay laboratory, has already been outlined above, and quartering as used in smelters. ffl

CONING AND QUARTERING

As the crushed ore is brought into the sampling room, it is evenly deposited on the sampling floor in a large ring. The man then shovels the ore into a conical heal in the centre of the floor, while walking slowly around the ring. Only a certain amount of the ore should be shovelled up in passing once around the heap, because if too much of it is shovelled up, a part of the ore may be too much bunched. He should throw each shovelful upon the apex of the cone, thus allowing each shovelful to be evenly distributed on all sides of the cone.

When the operation has been completed, the ore is raked out into another ring by means of a shovel or hoe. It can now be once more coned or it may be shovelled into a new cone on another part of the floor. This process of reconing is repeated until the ore is satisfactorily mixed, and then the ore is flattened down to a ring once more. This last flattening down of the cone is usually done in a systematic way by walking around the heap and raking it down. The ore, in the shape of a flattened cone, is now quartered with a stick along two diameters at right angles to each other. Two opposite quarters are shovelled away and the two remaining ones are again worked into a heap. The same operation is repeated until the sample is small enough, a recrushing being done between each operation, if necessary. One difficulty is that the fine material tends to separate out of the coarser substances, and thus hinder thorough mixing. If the ore is somewhat damp, this difficulty is partly overcome. It is also highly essential that the sampling floor shall be level, clean and without cracks, and for this reason sampling floors are usually covered with iron plates.

The split shovel is a fork in which the prongs form separate scoops, each scoop being the same size as the space between two scoops. The split shovel is laid down on the ground and the ore is spread evenly over its surface. It is now raised, and the ore lying between the scoops is left on the ground while that in the scoops is thrown in a heap by itself. By this means the ore can be halved as many times as is desired.

The riffle is simply a large split shovel which has a small handle on each side, instead of one handle at the back as in the case of the ordinary split shovel. The Jones sampler has two sets of scoops of the same hize, sloping in opposite directions, instead of alternate scoops and spaces as in the case of a split shovel or riffle.

Using a Jones sampler, the ore is discharged by the two sets of scoops just as fast as it is poured on to the sampler, and is deposited in two separate heaps on the sampling floor. In these devices, it is customary to have each scoop at least four times as wide as the largest particle of ore.

Besides these devices there are also pipe and grab samples. A pipe sampler is one obtained by driving a cheese scoop sampler or pipe into the ore. This can, of course, be done while the ore is still in the sacks, cars or bins. A grab sample is one taken by dividing the surface part of the ore into squares and taking equal quantities of ore from the corners of the squares. Running samples are also sometimes taken by hand devices, such as dipping a bucket into the stream at given intervals.

An assay of an ore is always made on a dry sample, because it is impossible to obtain accurate results on a wet sample. The ore, on the other hand, is usually always damp when samples, and it is for this reason that it becomes necessary to get "moisture samples," from which the percentage of foisture in the ore can be determined. The moisture sample is always taken just before or just after the ore is weighed and the method of obtaining such a moist sample must be a rapid one. The sample cannot be recrushed or cut down, as it has to be placed in a covered part immediately. By getting the weight of the moisture sample before any evaporation has taken place and its weight after it has been allowed to dry thoroughly and getting the difference of these two weights, one can determine how much moisture there was in the ore.

MECHANICAL SAMPLING

Some automatic samplers taken in part of the stream all of the time, whereas others take all of the stream part of the time. But as the values in an ore are never evenly distributed across the stream, the former is not so efficient a method as the latter. R. H. Richards mentions seven essential features of a perfect mechanical sampler, which are:

1. It must take the whole stream of ore (wet or dry) part of the time.

2. The scoop that cuts out the sample must move completely across and out of the stream in one direction at each cut, for, if it enters from one side, and is then withdrawn on the same side without having completely crossed the stream, more ore will be taken from the side at which the scoop enters and leaves than from the other side. He also adds that although such a scoop may take part of the stream at a given time, it does take a true section across the stream and virtually takes the whole stream part of the time.

3. The scoop must move at a uniform rate, and the top of the scoop must, in all positions, be at right angles to the direction of the stream, in order to take equal proportions from all parts of the stream. This condittion, he advises, is well obtained from a vertical stream and a horizontal scoop, in the case of a revolving scoop.

4. If the scoop that cuts out the sample revolves about an axis, two sides of the scoop should converge towards the

axis in order to take equal proportions from all parts of the stream and the scoop may be adjustable to take larger or smaller proportions of the ore.

5. The interval of time between cuts should be constant.

6. The scoops must be deep and broad enough so that ore that has once gotten into them will not bound out again; and if the scoops have closed bottoms they must not be allowed to fill up so that some of the ore runs over, as this would produce a concentration of the heavy minerals, especially when the ore is carried in running water.

7. The machine should be simple and easily accessible for cleaning, to avoid danger of contaminating subsequent samples.

There are some samplers now used in mills and smelters that cover all of these points mentioned by Dr. Richards in his "Text Book on Ore Dressing." Such samplers are the Snyder, the Vezin and the Brunton.

As hand sampling requires much more labor than mechanical sampling, it is more costly, and it is impossible to have an intentional error in the case of mechanical sampling.

The size of a sample depends largely on the character of the ore. If the metal values are very evenly distributed throughout the ore, a smaller sample is required than when this is not the case. From this it will be clear that the weight of the sample taken will decrease as the size of the ore particles decrease.

A considerable part of the ore is crushed fine when mechanical sampling is used, and since such fine material is undesirable in the blast furnace, lead smelters, which heat all of their ore by blast furnaces, still use hand sampling almost exclusively in the case of oxide ores. This, however, is not the case with the sulphide ores that come to a lead smelter, because they have to be roasted and sintered before smelting, and finely crushed ore facilities roasting. Thus it is customary in a lead smelter to use mechanical sampling on the sulphide ores and hand sampling on the oxide ores.

The following describes the method of sampling in use at the Tigre Mining Co., as described in the Engineering & Mining Journal of June 6, 1914. "The sorted high grade and sorted waste are grab-sampled; the ore on its way to the mill is sampled automatically in a revolving sampler of the slotted-cone type; the concentrate is sampled by the "cheese-trier" method; the stamp mill tailings is sampled by a Scobey-sampler and the dump tailings are uniform and sampled three times per shift by hand. The bullion is sampled by dipping; the precipitate by the same method as the concenerate, and the cyanide by hand, in the tailings sluices at each discharge of the filter press. Pregnant solutions by dropping a weighted bottle into the stump tanks, but barren solutions by drip methods."

The method of sampling at the Nipissing mill is as follows: The ore is automatically sampled on its way to the desulphurizing process. The pulp is sampled by cutting the stream every six minutes.

SAMPLING AT EL PASO SMELTERY

The first sampling is done by hand during the operation of unloading. Here each fifth or tenth shovel is thrown onto the sample heap. On the concentrates each fifth shovel is taken and this sample is halved. The shovel samples are taken to the sample mill, while the concentrate samples go to the quartering floor, where they are sampled according to the coning and quartering method. The final sample here weighs about 400 pounds.

The shovel samples from the ore lots are reduced in some automatic sampling and crushing equipment. There are four sets of such automatic sampling and crushing equipments, two for ores not exceeding three inches in size, and two for coarse material.

"The samples from the finishing rolls are dried in two stream drying rooms, and thence go to two backing rooms. The duplicate samples are handled in different rooms. These rooms have four Engelbach grinders and four Eaton pulp mixers. The latter are small revolving cylinders, having several interior shelves to assist in mixing the material. From the Eaton mixer the sample is riffled down on a Jones divider and then goes to the bucking board. From the bucking board the pulp returns to a small Eaton mixer and after being revolved for a few minutes, the material is poured out and the sample is placed in the pulp envelopes.

In the above stated cases, sampling methods have been outlined as used both in the mill and in the smelter. Hand sampling can be done very conveniently in both cases, and also mechanical, but there is a better chance to get running samples in a wet-concentrating plant than there is in the case of a smelter, where the ore usually comes in in the form of dry shipments. To what extent mechanical sampling should be carried on depends greatly upon the mechanical perfection of the samplers used, especially with respect to cleanliness and to their adaptability to make the loss, which is brought about by the extreme fineness of the material, as small as possible.

In conclusion it may be worth while to point out how impractical it is for a smelter to underpay any party from whom ore is bought, as some, who have small quantities of ore for sale, very often imagine. A smelter is always willing to present a seller with half of the sample obtained at the smelter, in order that he himself can have the sample assayed, and if his results should not agree with those obtained at the smelter, the latter is furthermore willing to refer the whole matter to an umpire assayer. The regular procedure of operations as carried out at a smelter makes it almost an impossibility to get such a false sample as to specially favor itself, and this, together with its willingness to have its own assay results confirmed by the seller, makes the practice of under paying the owners of ores very much impractical. From this one concludes that when a man has some ore which he wishes to sell to a smelter, the latter will almost without exception, give him full value for his property.

University of Arizona Bulletin

MINERAL TECHNOLOGY SERIES NO. 19   DECEMBER 10, 1917

SELECT BLOWPIPE AND ACID TESTS FOR MINERALS

ALUMINUM, Al.

1. Infusible aluminum minerals (also zinc silicates) ignited before and after adding cobalt nitrate solution give an intense blue color. Fusible minerals may give blue cobalt glass whether aluminum is present or not.

2. Ammonia gives a white gelatinous precipitate in solutions containing aluminum. Iron hydroxide, chromium hydroxide, calcium phosphate, calcium borate, and calcium fluorid are also precipitated by ammonia along with aluminum hydroxide.

AMMONIUM, NH_4.

1. Ammonium salts heated in the closed tube with potassium hydroxide or CaO (made by heating calcite) give the characteristic ammonia odor.

ANTIMONY, Sb.

1. Antimony minerals heated on the charcoal in the oxidizing flame give a white coating, antimony oxide, near the assay and dense white fumes without odor.

2. With bismuth flux on plaster, antimony compounds give a peach-red coating or an orange coating stippled with peach-red.

3. In the open tube antimony minerals give a non-volatile, amorphous, white sublimate on the under side of the tube.

4. Concentrated nitric acid oxides antimony sulphids and sulphosalts to antimony oxide, a white precipitate soluble in potassium hydroxide.

ARSENIC, As.

A.—Compounds Without Oxygen

1. On charcoal most arsenic minerals give a white volatile coating far from the assay and fumes with characteristic odor of arsine.

2. In the open tube minute, brilliant, colorless crystals.

3. In closed tube a black mirror of arsenic.

B.—Arsenates

4. Arsenates heated intensely in closed tube with charcoal give a black metallic mirror.

5. Nitric acid solutions of arsenates give a yellow precipitate with ammonium molybdate when heated to boiling.

BARIUM, Ba.

1. Yellowish-green flame (not made blue by hydrochloric acid).

2. Dilute sulphuric acid precipitates white barium sulphate from dilute solutions.

CHROMIUM, Cr.

1. The borax and sodium metaphosphate beads are emerald green in both oxidizing and reducing flames.

2. The sodium carbonate bead is yellow in oxidizing flame.

3. Chromate solutions give a dark red precipitate with silver nitrate.

COBALT, Co.

1. The borax and sodium metaphosphate beads are deep blue in both the oxidizing and reducing flames. This is a very delicate test.

2. Heated on charcoal in reducing flame, cobalt compounds become magnetic.

COPPER, Cu.

1. Green flame made azure-blue with hydrochloric acid.

2. Borax and sodium metaphosphate beads are blue in oxidizing flame and opaque red in reducing flame. In the presence of iron, the oxidizing flame bead is green or bluish-green.

3. On charcoal with soda in reducing flame and also with salt of phosphorous and metallic tin or charcoal, metallic copper (malleable) is obtained.

4. Solutions of copper minerals are blue (green in the presence of iron). Ammonia produces a deep blue coloration.

5. Copper solutions touched to a bright surface of iron, such as knife-blade or hammer, give a coating of metallic copper.

GOLD, Au.

1. With soda on charcoal, gold compounds give a malleable yellow button.

2. Gold may be identified in some of its rich ores by panning and washing away light quartz, rock, etc. Mercury is added to the concentrates. By grinding in a mortar an amalgam of gold is obtained. This may be heated on charcoal or in a closed tube and the mercury driven off. The residue is heated with a little borax on charcoal and a globule of gold obtained.

IRON, Fe.

1. On charcoal reducing flame, especially with soda, iron minerals become magnetic. (Also cobalt and nickel).

2. In oxidizing flame the borax bead is amber colored; in reducing flame, bottle green.

3. Ammonia precipitates brownish-red iron hydroxide. A few drops of nitric acid should always be added to the solution to insure oxidation of the iron.

4. To detect state of iron, a borax bead made blue with malachite is changed to opaque red by a ferrous compound, and to green by a ferric compound. (Use a neutral flame.)

LEAD, Pb.

1. On charcoal with soda in reducing flame a malleable button of lead and a yellow coating of lead oxide. Lead sulphide also gives a white coating of lead sulphate.

2. From solutions containing lead, hydrochloric acid precipitates lead chloride, which is soluble in hot water, but recrystallizes on cooling the solution, as white acicular crystals with adamantine lustre.

MAGNESIUM, Mg.

1. In the presence of ammonia and ammonium chloride, sodium phosphate precipitates ammonium magnesium phosphate, which forms slowly. Other metals (except alkalies) must be absent, as they also give precipitates.

MANGANESE, Mn.

1. Bluish-green soda bead, (a very delicate test.)

2. The borax or salt of phosphorous bead is amethyst colored in oxidizing flame and colorless in reducing flame.

3. With hydrochloric acid, manganese dioxids give off chlorine, a gas recognized by its penetrating odor.

MERCURY, Hg.

1. In closed tube with dry soda, gives globules of mercury.

2. On plaster with bismuth flux, a scarlet sublimate when gently heated. If overheated, the sublimate is dark greenish-yellow.

3. Most mercury compounds rubbed on a copper coin with hydrochloric acid give a white amalgam.

MOLYBDENUM, Mo.

1. Salt of phosphorous bead is green in reducing flame, but colorless in oxidizing flame. The reducing flame beads dissolved in hydrochloric acid with tin or paper, give a brown solution.

2. If a piece of paper is dropped into a concentrated sulphuric acid solution, it turns blue when hot, and on cooling becomes brown; on heating again, becomes blue.

NICKEL, Ni.

1. The borax bead in oxidizing flame is violet when hot, reddish-brown, cold; while in reducing flame the bead is turbid gray.

2. With nickel solutions sodium hydroxide gives a pale green precipitate which is insoluble in excess.

PHOSPHOROUS, P.

1. An excess of ammonium molybdate added to a nitric acid solution of a phosphate gives a yellow precipitate which is soluble in ammonia. The solution should be only slightly heated, for arsenates give a similar precipitate on boiling.

PLATINUM, Pt.

1. Insoluble in any single acid, but soluble in aqua regia (one third nitric and two thirds hydrochloric acid). In rather concentrated, slightly acid solutions, potassium chloride gives a yellow precipitate, platinic chloride, insoluble in alcohol.

SILVER, Ag.

1. With soda on charcoal in reducing flame, silver minerals yield malleable metallic globules of silver, which may be tested as under next test.

2. Nitric acid solutions of silver minerals on the addition of hydrochloric acid give a white curdy precipitate which changes to violet on exposure to light and is soluble in ammonia.

STRONTIUM, Sr.

1. Strontium compounds give a crimson flame, especially with hydrochloric acid.

TIN, Sn.

1. Tin compounds heated on charcoal in oxidizing flame give a straw-colored coating, tin oxide. On addition of cobalt nitrate solution and heating in reducing flame, a bluish-green coloration results.

2. Tin compounds fused on charcoal with soda and a little sulphur, in strong reducing flame, give malleable metallic buttons of tin, which are oxidized by nitric acid to a white insoluble powder.

TUNGSTEN, W.

1. The salt of phosphorous bead is blue in reducing flame, colorless in oxidizing flame. Iron interferes and gives a red bead in reducing flame.

2. Salt of phosphorous beads treated on charcoal in reducing flame with tin are dissolved in hydrochloric acid with the addition of metallic tin to a deep blue solution.

3. With soluble tungstates, hydrochloric acid gives a yellow residue, tungstic acid, which is soluble in ammonia.

VANADIUM, V.

1. The salt of phosphorous bead is a fine green in reducing flame, and light yellow in oxidizing flame.

2. In closed tube with potassium and sulphate, vanadates give a yellow mass.

ZINC, Zn.

1. On charcoal with soda, zinc compounds give a white coating which is yellow when hot.

2. Zinc minerals, when moistened with cobalt nitrate solution and intensely ignited, assume a bright green color, due to the formation of cobalt zincate. Zinc silicates give a blue color like aluminum compounds, but if tried on charcoal, the sublimate will turn green.

GOLD RUSH BOOKS

OREGON, USA

www.GoldMiningBooks.com

Books On Mining

Visit: www.goldminingbooks.com to order your copies or ask your favorite book seller to offer them.

Mining Books by Kerby Jackson

<u>Gold Dust: Stories From Oregon's Mining Years</u> - Oregon mining historian and prospector, Kerby Jackson, brings you a treasure trove of seventeen stories on Southern Oregon's rich history of gold prospecting, the prospectors and their discoveries, and the breathtaking areas they settled in and made homes. 5" X 8", 98 ppgs. Retail Price: $11.99

<u>The Golden Trail: More Stories From Oregon's Mining Years</u> - In his follow-up to "Gold Dust: Stories of Oregon's Mining Years", this time around, Jackson brings us twelve tales from Oregon's Gold Rush, including the story about the first gold strike on Canyon Creek in Grant County, about the old timers who found gold by the pail full at the Victor Mine near Galice, how Iradel Bray discovered a rich ledge of gold on the Coquille River during the height of the Rogue River War, a tale of two elderly miners on the hunt for a lost mine in the Cascade Mountains, details about the discovery of the famous Armstrong Nugget and others. 5" X 8", 70 ppgs. Retail Price: $10.99

Oregon Mining Books

<u>Geology and Mineral Resources of Josephine County, Oregon</u> - Unavailable since the 1970's, this important publication was originally compiled by the Oregon Department of Geology and Mineral Industries and includes important details on the economic geology and mineral resources of this important mining area in South Western Oregon. Included are notes on the history, geology and development of important mines, as well as insights into the mining of gold, copper, nickel, limestone, chromium and other minerals found in large quantities in Josephine County, Oregon. 8.5" X 11", 54 ppgs. Retail Price: $9.99

<u>Mines and Prospects of the Mount Reuben Mining District</u> - Unavailable since 1947, this important publication was originally compiled by geologist Elton Youngberg of the Oregon Department of Geology and Mineral Industries and includes detailed descriptions, histories and the geology of the Mount Reuben Mining District in Josephine County, Oregon. Included are notes on the history, geology, development and assay statistics, as well as underground maps of all the major mines and prospects in the vicinity of this much neglected mining district. 8.5" X 11", 48 ppgs. Retail Price: $9.99

<u>The Granite Mining District</u> - Notes on the history, geology and development of important mines in the well known Granite Mining District which is located in Grant County, Oregon. Some of the mines discussed include the Ajax, Blue Ribbon, Buffalo, Continental, Cougar-Independence, Magnolia, New York, Standard and the Tillicum. Also included are many rare maps pertaining to the mines in the area. 8.5" X 11", 48 ppgs. Retail Price: $9.99

<u>Ore Deposits of the Takilma and Waldo Mining Districts of Josephine County, Oregon</u> - The Waldo and Takilma mining districts are most notable for the fact that the earliest large scale mining of placer gold and copper in Oregon took place in these two areas. Included are details about some of the earliest large gold mines in the state such as the Llano de Oro, High Gravel, Cameron, Platerica, Deep Gravel and others, as well as copper mines such as the famous Queen of Bronze mine, the Waldo, Lily and Cowboy mines. This volume also includes six maps and 20 original illustrations. 8.5" X 11", 74 ppgs. Retail Price: $9.99

<u>Metal Mines of Douglas, Coos and Curry Counties, Oregon</u> - Oregon mining historian Kerby Jackson introduces us to a classic work on Oregon's mining history in this important re-issue of Bulletin 14C Volume 1, otherwise known as the Douglas, Coos & Curry Counties, Oregon Metal Mines Handbook. Unavailable since 1940, this important publication was originally compiled by the Oregon Department of Geology and Mineral Industries includes detailed descriptions, histories and the geology of over 250 metallic mineral mines and prospects in this rugged area of South West Oregon. 8.5" X 11", 158 ppgs. Retail Price: $19.99

Metal Mines of Jackson County, Oregon - Unavailable since 1943, this important publication was originally compiled by the Oregon Department of Geology and Mineral Industries includes detailed descriptions, histories and the geology of over 450 metallic mineral mines and prospects in Jackson County, Oregon. Included are such famous gold mining areas as Gold Hill, Jacksonville, Sterling and the Upper Applegate. **8.5" X 11", 220 ppgs. Retail Price: $24.99**

Metal Mines of Josephine County, Oregon - Oregon mining historian Kerby Jackson introduces us to a classic work on Oregon's mining history in this important re-issue of Bulletin 14C, otherwise known as the Josephine County, Oregon Metal Mines Handbook. Unavailable since 1952, this important publication was originally compiled by the Oregon Department of Geology and Mineral Industries includes detailed descriptions, histories and the geology of over 500 metallic mineral mines and prospects in Josephine County, Oregon. **8.5" X 11", 250 ppgs. Retail Price: $24.99**

Metal Mines of North East Oregon - Oregon mining historian Kerby Jackson introduces us to a classic work on Oregon's mining history in this important re-issue of Bulletin 14A and 14B, otherwise known as the North East Oregon Metal Mines Handbook. Unavailable since 1941, this important publication was originally compiled by the Oregon Department of Geology and Mineral Industries and includes detailed descriptions, histories and the geology of over 750 metallic mineral mines and prospects in North Eastern Oregon. **8.5" X 11", 310 ppgs. Retail Price: $29.99**

Metal Mines of North West Oregon - Oregon mining historian Kerby Jackson introduces us to a classic work on Oregon's mining history in this important re-issue of Bulletin 14D, otherwise known as the North West Oregon Metal Mines Handbook. Unavailable since 1951, this important publication was originally compiled by the Oregon Department of Geology and Mineral Industries and includes detailed descriptions, histories and the geology of over 250 metallic mineral mines and prospects in North Western Oregon. **8.5" X 11", 182 ppgs. Retail Price: $19.99**

Mines and Prospects of Oregon - Mining historian Kerby Jackson introduces us to a classic mining work by the Oregon Bureau of Mines in this important re-issue of The Handbook of Mines and Prospects of Oregon. Unavailable since 1916, this publication includes important insights into hundreds of gold, silver, copper, coal, limestone and other mines that operated in the State of Oregon around the turn of the 19th Century. Included are not only geological details on early mines throughout Oregon, but also insights into their history, production, locations and in some cases, also included are rare maps of their underground workings. **8.5" X 11", 314 ppgs. Retail Price: $24.99**

Lode Gold of the Klamath Mountains of Northern California and South West Oregon
(See California Mining Books)

Mineral Resources of South West Oregon - Unavailable since 1914, this publication includes important insights into dozens of mines that once operated in South West Oregon, including the famous gold fields of Josephine and Jackson Counties, as well as the Coal Mines of Coos County. Included are not only geological details on early mines throughout South West Oregon, but also insights into their history, production and locations. **8.5" X 11", 154 ppgs. Retail Price: $11.99**

Chromite Mining in The Klamath Mountains of California and Oregon
(See California Mining Books)

Southern Oregon Mineral Wealth - Unavailable since 1904, this rare publication provides a unique snapshot into the mines that were operating in the area at the time. Included are not only geological details on early mines throughout South West Oregon, but also insights into their history, production and locations. Some of the mining areas include Grave Creek, Greenback, Wolf Creek, Jump Off Joe Creek, Granite Hill, Galice, Mount Reuben, Gold Hill, Galls Creek, Kane Creek, Sardine Creek, Birdseye Creek, Evans Creek, Foots Creek, Jacksonville, Ashland, the Applegate River, Waldo, Kerby and the Illinois River, Althouse and Sucker Creek, as well as insights into local copper mining and other topics. **8.5" X 11", 64 ppgs. Retail Price: $8.99**

Geology and Ore Deposits of the Takilma and Waldo Mining Districts - Unavailable since the 1933, this publication was originally compiled by the United States Geological Survey and includes details on gold and copper mining in the Takilma and Waldo Districts of Josephine County, Oregon. The Waldo and Takilma mining districts are most notable for the fact that the earliest large scale mining of placer gold and copper in Oregon took place in these two areas. Included in this report are details about some of the earliest large gold mines in the state such as the Llano de Oro, High Gravel, Cameron, Platerica, Deep Gravel and others, as well as copper mines such as the famous Queen of Bronze mine, the Waldo, Lily and Cowboy mines. In addition to geological examinations, insights are also provided into the production, day to day operations and early histories of these mines, as well as calculations of known mineral reserves in the area. This volume also includes six maps and 20 original illustrations. **8.5" X 11", 74 ppgs. Retail Price: $9.99**

Gold Mines of Oregon - Oregon mining historian Kerby Jackson introduces us to a classic work on Oregon's mining history in this important re-issue of Bulletin 61, otherwise known as "Gold and Silver In Oregon". Unavailable since 1968, this important publication was originally compiled by geologists Howard C. Brooks and Len Ramp of the Oregon Department of Geology and Mineral Industries and includes detailed descriptions, histories and the geology of over 450 gold mines Oregon. Included are notes on the history, geology and gold production statistics of all the major mining areas in Oregon including the Klamath Mountains, the Blue Mountains and the North Cascades. While gold is where you find it, as every miner knows, the path to success is to prospect for gold where it was previously found. **8.5" X 11", 344 ppgs. Retail Price: $24.99**

Mines and Mineral Resources of Curry County Oregon - Originally published in 1916, this important publication on Oregon Mining has not been available for nearly a century. Included are rare insights into the history, production and locations of dozens of gold mines in Curry County, Oregon, as well as detailed information on important Oregon mining districts in that area such as those at Agness, Bald Face Creek, Mule Creek, Boulder Creek, China Diggings, Collier Creek, Elk River, Gold Beach, Rock Creek, Sixes River and elsewhere. Particular attention is especially paid to the famous beach gold deposits of this portion of the Oregon Coast. **8.5" X 11", 140 ppgs. Retail Price: $11.99**

Chromite Mining in South West Oregon - Originally published in 1961, this important publication on Oregon Mining has not been available for nearly a century. Included are rare insights into the history, production and locations of nearly 300 chromite mines in South Western Oregon. **8.5" X 11", 184 ppgs. Retail Price: $14.99**

Mineral Resources of Douglas County Oregon - Originally published in 1972, this important publication on Oregon Mining has not been available for nearly forty years. Included are rare insights into the geology, history, production and locations of numerous gold mines and other mining properties in Douglas County, Oregon. **8.5" X 11", 124 ppgs. Retail Price: $11.99**

Mineral Resources of Coos County Oregon - Originally published in 1972, this important publication on Oregon Mining has not been available for nearly forty years. Included are rare insights into the geology, history, production and locations of numerous gold mines and other mining properties in Coos County, Oregon. **8.5" X 11", 100 ppgs. Retail Price: $11.99**

Mineral Resources of Lane County Oregon - Originally published in 1938, this important publication on Oregon Mining has not been available for nearly seventy five years. Included are extremely rare insights into the geology and mines of Lane County, Oregon, in particular in the Bohemia, Blue River, Oakridge, Black Butte and Winberry Mining Districts. **8.5" X 11", 82 ppgs. Retail Price: $9.99**

Mineral Resources of the Upper Chetco River of Oregon: Including the Kalmiopsis Wilderness - Originally published in 1975, this important publication on Oregon Mining has not been available for nearly forty years. Withdrawn under the 1872 Mining Act since 1984, real insight into the minerals resources and mines of the Upper Chetco River has long been unavailable due to the remoteness of the area. Despite this, the decades of battle between property owners and environmental extremists over the last private mining inholding in the area has continued to pique the interest of those interested in mining and other forms of natural resource use. Gold mining began in the area in the 1850's and has a rich history in this geographic area, even if the facts surrounding it are little known. Included are twenty two rare photographs, as well as insights into the Becca and Morning Mine, the Emmly Mine (also known as Emily Camp), the Frazier Mine, the Golden Dream or Higgins Mine, Hustis Mine, Peck Mine and others. **8.5" X 11", 64 ppgs. Retail Price: $8.99**

Gold Dredging in Oregon - Originally published in 1939, this important publication on Oregon Mining has not been available for nearly seventy five years. Included are extremely rare insights into the history and day to day operations of the dragline and bucketline gold dredges that once worked the placer gold fields of South West and North East Oregon in decades gone by. Also included are details into the areas that were worked by gold dredges in Josephine, Jackson, Baker and Grant counties, as well as the economic factors that impacted this mining method. This volume also offers a unique look into the values of river bottom land in relation to both farming and mining, in how farm lands were mined, re-soiled and reclamated after the dredges worked them. Featured are hard to find maps of the gold dredge fields, as well as rare photographs from a bygone era. **8.5" X 11", 86 ppgs. Retail Price: $8.99**

Quick Silver Mining in Oregon - Originally published in 1963, this important publication on Oregon Mining has not been available for over fifty years. This publication includes details into the history and production of Elemental Mercury or Quicksilver in the State of Oregon. **8.5" X 11", 238 ppgs. Retail Price: $15.99**

Mines of the Greenhorn Mining District of Grant County Oregon - Originally published in 1948, this important publication on Oregon Mining has not been available for over sixty five years. In this publication are rare insights into the mines of the famous Greenhorn Mining District of Grant County, Oregon, especially the famous Morning Mine. Also included are details on the Tempest, Tiger, Bi-Metallic, Windsor, Psyche, Big Johnny, Snow Creek, Banzette and Paramount Mines, as well as prospects in the vicinities in the famous mining areas of Mormon Basin, Vinegar Basin and Desolation Creek. Included are hard to find mine maps and dozens of rare photographs from the bygone era of Grant County's rich mining history. **8.5" X 11", 72 ppgs. Retail Price: $9.99**

Geology of the Wallowa Mountains of Oregon: Part I (Volume 1) - Originally published in 1938, this important publication on Oregon Mining has not been available for nearly seventy five years. Included are details on the geology of this unique portion of North Eastern Oregon. This is the first part of a two book series on the area. Accompanying the text are rare photographs and historic maps.8.5" X 11", 92 ppgs. Retail Price: $9.99

Geology of the Wallowa Mountains of Oregon: Part II (Volume 2) - Originally published in 1938, this important publication on Oregon Mining has not been available for nearly seventy five years. Included are details on the geology of this unique portion of North Eastern Oregon. This is the first part of a two book series on the area. Accompanying the text are rare photographs and historic maps.8.5" X 11", 94 ppgs. Retail Price: $9.99

Field Identification of Minerals For Oregon Prospectors - Originally published in 1940, this important publication on Oregon Mining has not been available for nearly seventy five years. Included in this volume is an easy system for testing and identifying a wide range of minerals that might be found by prospectors, geologists and rockhounds in the State of Oregon, as well as in other locales. Topics include how to put together your own field testing kit and how to conduct rudimentary tests in the field. This volume is written in a clear and concise way to make it useful even for beginners. 8.5" X 11", 158 ppgs. Retail Price: $14.99

The Bohemia Mining District of Oregon - Originally published in 1900, this important publication on Oregon Mining has not been available for over a century. Included in this volume are important insights into the famous Bohemia Mining District of Oregon, including the histories and locations of important gold mines in the area such as the Ophir Mine, Clarence, Acturas, Peek-a-boo, White Swan, Combination Mine, the Musick Mine, The California, White Ghost, The Mystery, Wall Street, Vesuvius, Story, Lizzie Bullock, Delta, Elsie Dora, Golden Slipper, Broadway, Champion Mine, Knott, Noonday, Helena, White Wings, Riverside and others. Also included are notes on the nearby Blue River Mining District. 8.5" X 11", 58 ppgs. Retail Price: $9.99

The Gold Fields of Eastern Oregon - Unavailable since 1900, this publication was originally compiled by the Baker City Chamber of Commerce Offering important insights into the gold mining history of Eastern Oregon, "The Gold Fields of Eastern Oregon" sheds a rare light on many of the gold mines that were operating at the turn of the 19th Century in Baker County and Grant County in North Eastern Oregon. Some of the areas featured include the Cable Cove District, Baisely-Elhorn, Granite, Red Boy, Bonanza, Susanville, Sparta, Virtue, Vaughn, Sumpter, Burnt River, Rye Valley and other mining districts. Included is basic information on not only many gold mines that are well known to those interested in Eastern Oregon mining history, but also many mines and prospects which have been mostly lost to the passage of time. Accompanying are numerous rare photos 8.5" X 11", 78 ppgs. Retail Price: $10.99

Gold Mining in Eastern Oregon - Originally published in 1938, this important publication on Oregon Mining has not been available for over a century. Included in this volume are important insights into the famous mining districts of Eastern Oregon during the late 1930's. Particular attention is given to those gold mines with milling and concentrating facilities in the Greenhorn, Red Boy, Alamo, Bonanza, Granite, Cable Cove, Cracker Creek, Virtue, Keating, Medical Springs, Sanger, Sparta, Chicken Creek, Mormon Basin, Connor Creek, Cornucopia and the Bull Run Mining Districts. Some of the mines featured include the Ben Harrison, North Pole-Columbia, Highland Maxwell, Baisley-Elkhorn, White Swan, Balm Creek, Twin Baby, Gem of Sparta, New Deal, Gleason, Gifford-Johnson, Cornucopia, Record, Bull Run, Orion and others. Of particular interest are the mill flow sheets and descriptions of milling operations of these mines. 8.5" X 11", 68 ppgs. Retail Price: $8.99

The Gold Belt of the Blue Mountains of Oregon - Originally published in 1901, this important publication on Oregon Mining has not been available for over a century. Included in this volume are rare insights into the gold deposits of the Blue Mountains of North East Oregon, including the history of their early discovery and early production. Extensive details are offered on this important mining area's mineralogy and economic geology, as well as insights into nearby gold placers, silver deposits and copper deposits. Featured are the Elkhorn and Rock Creek mining districts, the Pocahontas district, Auburn and Minersville districts, Sumpter and Cracker Creek, Cable Cove, the Camp Carson district, Granite, Alamo, Greenhorn, Robinsonville, the Upper Burnt River Valley and Bonanza districts, Susanville, Quartzburg, Canyon Creek, Virtue, the Copper Butte district, the North Powder River, Sparta, Eagle Creek, Cornucopia, Pine Creek, Lower Powder River, the Upper Snake River Canyon, Rye Valley, Lower Burnt River Valley, Mormon Basin, the Malheur and Clarks Creek districts, Sutton Creek and others. Of particular interest are important details on numerous gold mines and prospects in these mining districts, including their locations, histories, geology and other important information, as well as information on silver, copper and fire opal deposits. 8.5" X 11", 250 ppgs. Retail Price: $24.99

Mining in the Cascades Range of Oregon - Originally published in 1938, this important publication on Oregon Mining has not been available for over seventy five years. Included in this volume are rare insights into the gold mines and other types of metal mines in the Cascades Mountain Range of Oregon. Some of the important mining areas covered include the famous Bohemia Mining District, the North Santiam Mining District, Quartzville Mining District, Blue River Mining District, Fall Creek Mining District, Oakridge District, Zinc District, Buzzard-Al Sarena District, Grand Cove, Climax District and Barron Mining District. Of particular interest are important details on over 100 mines and prospects in these mining districts, including their locations, histories, geology and other important information. 8.5" X 11", 170 ppgs. Retail Price: $14.99

Beach Gold Placers of the Oregon Coast - Originally published in 1934, this important publication on Oregon Mining has not been available for over 80 years. Included in this volume are rare insights into the beach gold deposits of the State of Oregon, including their locations, occurance, composition and geology. Of particular interest is information on placer platinum in Oregon's rich beach deposits. Also included are the locations and other information on some famous Oregon beach mines, including the Pioneer, Eagle, Chickamin, Iowa and beach placer mines north of the mouth of the Rogue River. 8.5" X 11", 60 ppgs. Retail Price: $8.99

Mineralogical Composition of the Sands of the Oregon Coast: From Coos Bay to the Columbia - Published in 1945, he text features hard to find information on the composition of the gold bearing black sands of the South West Oregon Coast, offering a unique insight to prospectors in search of Oregon's legendary beach gold. 104 ppgs, $9.99

Manganese Mining in Oregon - First released in 1942 and now out of print, this special reprint edition of "Manganese in Oregon" was originally published by the Oregon Department of Geology and Mineral Industries. The text features hard to find information on the mining of Manganese in Oregon, including details and maps of Oregon manganese mines and prospects. 108 ppgs, 9.99

Medford Oregon As A Mining Center - Written in 1912, this hard to find publication includes valuable insights into the mining history of South West Oregon. This small book contains interesting information on the gold, copper and mining industry in Southern Oregon as it existed just prior to World War One, shedding light on some of the important mines in the area. Included are rare photographs and vintage advertising of the day. 80 ppgs, 9.99

Mineral Resources of Curry County Oregon - First released in 1977 and now out of print, this special reprint edition of "Geology, Mineral Resources and Rock Materials of Curry County, Oregon" was originally published in cooperation of Curry County, Oregon and the Oregon Department of Geology and Mineral Industries. The text features hard to find information on not only the mining of gold and other metals in Curry County, but also aggregate mining in the area. 102 ppgs, 11.99

Origin of the Gold Bearing Black Sands of the Coast of South West Oregon - First released in 1943 and now out of print, this special reprint edition of "The Origin of the Black Sands of the South West Oregon Coast" was originally published by the Oregon Department of Geology and Mineral Industries. The text features hard to find information on the origin of the gold bearing black sands of the South West Oregon Coast, offering a unique insight to prospectors in search of Oregon's legendary beach gold. 52 ppgs, 8.99

South West Oregon Mining - Leading mining historian Kerby Jackson introduces us to six classic small mining publications on the Gold Mining Industry in Southern Oregon. This small book consists of a compilation of USGS J.S. Diller's "Mines of the Riddles Quadrangle", "The Rogue River Valley Coal Fields" and "Mineral Resources of the Grants Pass Quadrangle", the Grants Pass Commercial Club's rare publication "Mining in Josephine County, Oregon" and the USGS publication "The Distribution of Placer Gold in the Sixes River, South West Oregon". Also included is F.W. Libbey's legendary article on the Southern Oregon Mining Industry, "Lest We Forget", which appeared in the publication of the Oregon State Department of Geology and Mineral Industries in the early 1960's. This compilation offers a unique perspective on mining in South West Oregon and includes considerable information on mines in Josephine, Jackson and Coos Counties. 142 ppgs, 14.99

Geology and Mineral Resources of the Gasquet Quadrangle of California-Oregon - First published in 1953, it has been unavailable for over a century and sheds important light on the geological features and mineral resources of this portion of Northern California and Southern Oregon. 80 ppgs, 9.99

Idaho Mining Books

Gold in Idaho - Unavailable since the 1940's, this publication was originally compiled by the Idaho Bureau of Mines and includes details on gold mining in Idaho. Included is not only raw data on gold production in Idaho, but also valuable insight into where gold may be found in Idaho, as well as practical information on the gold bearing rocks and other geological features that will assist those looking for placer and lode gold in the State of Idaho. This volume also includes thirteen gold maps that greatly enhance the practical usability of the information contained in this small book detailing where to find gold in Idaho. **8.5" X 11", 72 ppgs. Retail Price: $9.99**

Geology of the Couer D'Alene Mining District of Idaho - Unavailable since 1961, this publication was originally compiled by the Idaho Bureau of Mines and Geology and includes details on the mining of gold, silver and other minerals in the famous Coeur D'Alene Mining District in Northern Idaho. Included are details on the early history of the Coeur D'Alene Mining District, local tectonic settings, ore deposit features, information on the mineral belts of the Osburn Fault, as well as detailed information on the famous Bunker Hill Mine, the Dayrock Mine, Galena Mine, Lucky Friday Mine and the infamous Sunshine Mine. This volume also includes sixteen hard to find maps. **8.5" X 11", 70 ppgs. Retail Price: $9.99**

The Gold Camps and Silver Cities of Idaho - Originally published in 1963, this important publication on Idaho Mining has not been available for nearly fifty years. Included are rare insights into the history of Idaho's Gold Rush, as well as the mad craze for silver in the Idaho Panhandle. Documented in fine detail are the early mining excitements at Boise Basin, at South Boise, in the Owyhees, at Deadwood, Long Valley, Stanley Basin and Robinson Bar, at Atlanta, on the famous Boise River, Volcano, Little Smokey, Banner, Boise Ridge, Hailey, Leesburg, Lemhi, Pearl, at South Mountain, Shoup and Ulysses, Yellow Jacket and Loon Creek. The story follows with the appearance of Chinese miners at the new mining camps on the Snake River, Black Pine, Yankee Fork, Bay Horse, Clayton, Heath, Seven Devils, Gibbonsville, Vienna and Sawtooth City. Also included are special sections on the Idaho Lead and Silver mines of the late 1800's, as well as the mining discoveries of the early 1900's that paved the way for Idaho's modern mining and mineral industry. Lavishly illustrated with rare historic photos, this volume provides a one of a kind documentary into Idaho's mining history that is sure to be enjoyed by not only modern miners and prospectors who still scour the hills in search of nature's treasures, but also those enjoy history and tromping through overgrown ghost towns and long abandoned mining camps. **8.5" X 11", 186 ppgs. Retail Price: $14.99**

Ore Deposits and Mining in North Western Custer County Idaho - Unavailable since 1913, this important publication was originally published by the Us Department of the Interior and has been unavailable for a century. Included are fine details on the geology, geography, gold placers and gold and silver bearing quartz veins of the mining region of North West Custer County, Idaho. Of particular interest is a rare look at the mines and prospects of the region, including those such as the Ramshorn Mine, SkyLark, Riverview, Excelsior, Beardsley, Pacific, Hoosier, Silver Brick, Forest Rose and dozens of others in the Bay Horse Mining District. Also covered are the mines of the Yankee Fork District such as the Lucky Boy, Badger, Black, Enterprise, Charles Dickens, Morrison, Golden Sunbeam, Montana, Golden Gate and others, as well as those in the Loon Mining District. **8.5" X 11", 126 ppgs. Retail Price: $12.99**

Gold Rush To Idaho - Unavailable since 1963, this important publication was originally published by the Idaho Bureau of Mines and has been unavailable for 50 years. "Gold Rush To Idaho" revisits the earliest years of the discovery of gold in Idaho Territory and introduces us to the conditions that the pioneer gold seekers met when they blazed a trail through the wilderness of Idaho's mountains and discovered the precious yellow metal at Oro Fino and Pierce. Subsequent rushes followed at places like Elk City, Newsome, Clearwater Station, Florence, Warrens and elsewhere. Of particular interest is a rare look at the hardships that the first miners in Idaho met with during their day to day existences and their attempts to bring law and order to their mining camps. **8.5" X 11", 88 ppgs. Retail Price: $9.99**

The Geology and Mines of Northern Idaho and North Western Montana - Unavailable since 1909, this important publication was originally published by the Us Department of the Interior and has been unavailable for a century. Included are fine details on the geology and geography of the mining regions of Northern Idaho and North Western Montana. Of particular interest is a rare look at the mines and prospects of the region, including those in the Pine Creek Mining District, Lake Pend Oreille district, Troy Mining District, Sylvanite District, Cabinet Mining District, Prospect Mining District and the Missoula Valley. Some of the mines featured include the Iron Mountain, Silver Butte, Snowshoe, Grouse Mountain Mine and others. **8.5" X 11", 142 ppgs. Retail Price: $12.99**

Mining in the Alturas Quadrangle of Blaine County Idaho - Unavailable since 1922, this important publication was originally published by the Idaho Bureau of Mines and has been unavailable for ninety years. Topics include the geology, rock formations and the formation of ore deposits in this important mining area of Idaho. Of particular focus is information on the local geology, quartz veins and ore deposits of this portion of Idaho. Included are hard to find details, including the descriptions and locations of numerous gold and silver mines in the area including the Silver King, Pilgrim, Columbia, Lone Jack, Sunbeam, Pride of the West, Lucky Boy, Scotia, Atlanta, Beaver-Bidwell and others mines and prospects. **8.5" X 11", 56 ppgs. Retail Price: $8.99**

Mining in Lemhi County Idaho - Originally published in 1913, this important book on Idaho Mining has not been available to miners for over a century. Included are rare insights into hundreds of gold, silver, copper and other mines in this famous Idaho mining area. Details include the locations, geology, history, production and other facts of the mines of this region, not only gold and silver hardrock mines, but also gold placer mines, lead-silver deposits, copper mines, cobalt-nickel deposits, tungsten and tin mines . It is lavishly illustrated with hard to find photos of the period and rare mining maps. Some of the vicinities featured include the Nicholia Mining District, Spring Mountain District, Texas District, Blue Wing District, Junction District, McDevitt District, Pratt Creek, Eldorado District, Kirtley Creek, Carmen Creek, Gibbonsville, Indian Creek, Mineral Hill District, Mackinaw, Eureka District, Blackbird District, YellowJacket District, Gravel Range District, Junction District, Parker Mountain and other mining districts. **8.5″ X 11″, 226 ppgs. Retail Price: $19.99**

Mining in Shoshone County Idaho - First published in 1923, it has been unavailable for over a century and sheds important light on the mining history of Shoshone County, Idaho. Some of the topics include the history of mining in Shoshone County, a look at the local geology and ore characteristics of lead-silver deposits, zinc deposits, copper, antimony, gold and other minerals. Also included are insights into the history, production, characteristics and locations of numerous mines in the area. 198 ppgs, 15.99

Utah Mining Books

Fluorite in Utah - Unavailable since 1954, this publication was originally compiled by the USGS, State of Utah and U.S. Atomic Energy Commission and details the mining of fluorspar, also known as fluorite in the State of Utah. Included are details on the geology and history of fluorspar (fluorite) mining in Utah, including details on where this unique gem mineral may be found in the State of Utah. **8.5″ X 11″, 60 ppgs. Retail Price: $8.99**

The Gold Hill Mining District of Utah - First published in 1935, it has been unavailable since those days and sheds important light on the mines, history and geology of Utah's Gold Hill Mining District. Included are rare insights into this important mining area, including the locations, histories and details of numerous mines. This volume is well illustrated with geological diagrams, as well as hard to find maps of some of the most important mines in this district. 202 ppgs., 19.99

The Mines, Miners and Minerals of Utah - First published in 1896, it has been unavailable since those days and sheds important light on the early mines and miners of Pioneer Utah, as well as the minerals which they won from the earth by laborious hard physical labor and sheer determination. Included are rare insights into the early mining history of Utah, as well details on hundreds of gold, silver and copper mines. 376 ppgs., 24.99

California Mining Books

The Tertiary Gravels of the Sierra Nevada of California - Mining historian Kerby Jackson introduces us to a classic mining work by Waldemar Lindgren in this important re-issue of The Tertiary Gravels of the Sierra Nevada of California. Unavailable since 1911, this publication includes details on the gold bearing ancient river channels of the famous Sierra Nevada region of California. **8.5″ X 11″, 282 ppgs. Retail Price: $19.99**

The Mother Lode Mining Region of California - Unavailable since 1900, this publication includes details on the gold mines of California's famous Mother Lode gold mining area. Included are details on the geology, history and important gold mines of the region, as well as insights into historic mining methods, mine timbering, mining machinery, mining bell signals and other details on how these mines operated. Also included are insights into the gold mines of the California Mother Lode that were in operation during the first sixty years of California's mining history. **8.5″ X 11″, 176 ppgs. Retail Price: $14.99**

Lode Gold of the Klamath Mountains of Northern California and South West Oregon - Unavailable since 1971, this publication was originally compiled by Preston E. Hotz and includes details on the lode mining districts of Oregon and California's Klamath Mountains. Included are details on the geology, history and important lode mines of the French Gulch, Deadwood, Whiskeytown, Shasta, Redding, Muletown, South Fork, Old Diggings, Dog Creek (Delta), Bully Choop (Indian Creek), Harrison Gulch, Hayfork, Minersville, Trinity Center, Canyon Creek, East Fork, New River, Denny, Liberty (Black Bear), Cecilville, Callahan, Yreka, Fort Jones and Happy Camp mining districts in California, as well as the Ashland, Rogue River, Applegate, Illinois River, Takilma, Greenback, Galice, Silver Peak, Myrtle Creek and Mule Creek districts of South Western Oregon. Also included are insights into the mineralization and other characteristics of this important mining region. **8.5″ X 11″, 100 ppgs. Retail Price: $10.99**

Mines and Mineral Resources of Shasta County, Siskiyou County, Trinity County: California - Unavailable since 1915, this publication was originally compiled by the California State Mining Bureau and includes details on the gold mines of this area of Northern California. Also included are insights into the mineralization and other characteristics of this important mining region, as well as the location of historic gold mines. **8.5″ X 11″, 204 ppgs. Retail Price: $19.99**

Geology of the Yreka Quadrangle, Siskiyou County, California - Unavailable since 1977, this publication was originally compiled by Preston E. Hotz and includes details on the geology of the Yreka Quadrangle of Siskiyou County, California. Also included are insights into the mineralization and other characteristics of this important mining region. **8.5" X 11", 78 ppgs. Retail Price: $7.99**

Mines of San Diego and Imperial Counties, California - Originally published in 1914, this important publication on California Mining has not been available for a century. This publication includes important information on the early gold mines of San Diego and Imperial County, which were some of the first gold fields mined in California by early Spanish and Mexican miners before the 49ers came on the scene. Included are not only details on early mining methods in the area, production statistics and geological information, but also the location of the early gold mines that helped make California "The Golden State". Also included are details on the mining of other minerals such as silver, lead, zinc, manganese, tungsten, vanadium, asbestos, barite, borax, cement, clay, dolomite, fluospar, gem stones, graphite, marble, salines, petroleum, stronium, talc and others. **8.5" X 11", 116 ppgs. Retail Price: $12.99**

Mines of Sierra County, California - Unavailable since 1920, this publication was originally compiled by the California State Mining Bureau and includes details on the gold mines of Sierra County, California. Also included are insights into the mineralization and other characteristics of this important mining region, as well as the location of historic gold mines. **8.5" X 11", 156 ppgs. Retail Price: $19.99**

Mines of Plumas County, California - Unavailable since 1918, this publication was originally compiled by the California State Mining Bureau and includes details on the gold mines of Plumas County, California. Also included are insights into the mineralization and other characteristics of this important mining region, as well as the location of historic gold mines. **8.5" X 11", 200 ppgs. Retail Price: $19.99**

Mines of El Dorado, Placer, Sacramento and Yuba Counties, California - Originally published in 1917, this important publication on California Mining has not been available for nearly a century. This publication includes important information on the early gold mines of El Dorado County, Placer County, Sacramento County and Yuba County, which were some of the first gold fields mined by the Forty-Niners during the California Gold Rush. Included are not only details on early mining methods in the area, production statistics and geological information, but also the location of the early gold mines that helped make California "The Golden State". Also included are insights into the early mining of chrome, copper and other minerals in this important mining area. **8.5" X 11", 204 ppgs. Retail Price: $19.99**

Mines of Los Angeles, Orange and Riverside Counties, California - Originally published in 1917, this important publication on California Mining has not been available for nearly a century. This publication includes important information on the early gold mines of Los Angeles County, Orange County and Riverside County, which were some of the first gold fields mined in California by early Spanish and Mexican miners before the 49ers came on the scene. Included are not only details on early mining methods in the area, production statistics and geological information, but also the location of the early gold mines that helped make California "The Golden State". **8.5" X 11", 146 ppgs. Retail Price: $12.99**

Mines of San Bernadino and Tulare Counties, California - Originally published in 1917, this important publication on California Mining has not been available for nearly a century. This publication includes important information on the early gold mines of San Bernadino and Tulare County, which were some of the first gold fields mined in California by early Spanish and Mexican miners before the 49ers came on the scene. Included are not only details on early mining methods in the area, production statistics and geological information, but also the location of the early gold mines that helped make California "The Golden State". Also included are details on the mining of other minerals such as copper, iron, lead, zinc, manganese, tungsten, vanadium, asbestos, barite, borax, cement, clay, dolomite, fluospar, gem stones, graphite, marble, salines, petroleum, stronium, talc and others. **8.5" X 11", 200 ppgs. Retail Price: $19.99**

Chromite Mining in The Klamath Mountains of California and Oregon - Unavailable since 1919, this publication was originally compiled by J.S. Diller of the United States Department of Geological Survey and includes details on the chromite mines of this area of Northern California and Southern Oregon. Also included are insights into the mineralization and other characteristics of this important mining region, as well as the location of historic mines. Also included are insights into chromite mining in Eastern Oregon and Montana. **8.5" X 11", 98 ppgs. Retail Price: $9.99**

Mines and Mining in Amador, Calaveras and Tuolumne Counties, California - Unavailable since 1915, this publication was originally compiled by William Tucker and includes details on the mines and mineral resources of this important California mining area. Included are details on the geology, history and important gold mines of the region, as well as insights into other local mineral resources such as asbestos, clay, copper, talc, limestone and others. Also included are insights into the mineralization and other characteristics of this important portion of California's Mother Lode mining region. **8.5" X 11", 198 ppgs. Retail Price: $14.99**

The Cerro Gordo Mining District of Inyo County California - Unavailable since 1963, this publication was originally compiled by the United States Department of Interior. Included are insights into the mineralization and other characteristics of this important mining region of Southern California. Topics include the mining of gold and silver in this important mining district in Inyo County, California, including details on the history, production and locations of the Cerro Gordo Mine, the Morning Star Mine, Estelle Tunnel, Charles Lease Tunnel, Ignacio, Hart, Crosscut Tunnel, Sunset, Upper Newtown, Newtown, Ella, Perseverance, Newsboy, Belmont and other silver and gold mines in the Cerro Gordo Mining District. This volume also includes important insights into the fossil record, geologic formations, faults and other aspects of economic geology in this California mining district. **8.5" X 11", 104 ppgs. Retail Price: $10.99**

Mining in Butte, Lassen, Modoc, Sutter and Tehama Counties of California - Unavailable since 1917, this publication was originally compiled by the United States Department of Interior. Included are insights into the mineralization and other characteristics of this important mining region of California. Topics include the mining of asbestos, chromite, gold, diamonds and manganese in Butte County, the mining of gold and copper in the Hayden Hill and Diamond Mountain mining districts of Lassen County, the mining of coal, salt, copper and gold in the High Grade and Winters mining districts of Modoc County, gold mining in Sutter County and the mining of gold, chromite, manganese and copper in Tehama County. This volume also includes the production records and locations of numerous mines in this important mining region. **8.5" X 11", 114 ppgs. Retail Price: $11.99**

Mines of Trinity County California - Originally published in 1965, this important publication on California Mining has not been available for nearly fifty years. This publication includes important information on mines and mining in Trinity County, California, as well insights into the mineralization and geology of this important mining area in Northern California. Included are extensive details on hardrock and placer gold mines and prospects, including charts showing the locations of these historic mines.. **8.5" X 11", 144 ppgs. Retail Price: $12.99**

Mines of Kern County California - Originally published in 1962, this important publication on California Mining has not been available for nearly fifty years. This publication includes important information on mines and mining in Kern County, California, as well insights into the mineralization and geology of this important mining area in California. Included are extensive details on hardrock and placer gold mines and prospects, including charts showing the locations of these historic mines. **8.5" X 11", 398 ppgs. Retail Price: $24.99**

Mines of Calaveras County California - Originally published in 1962, this important publication on California Mining has not been available for nearly fifty years. This publication includes important information on mines and mining in Calaveras County, California, as well insights into the mineralization and geology of this important mining area in Northern California. Included are extensive details on hardrock and placer gold mines and prospects, including charts showing the locations of these historic mines. **8.5" X 11", 236 ppgs. Retail Price: $19.99**

Lode Gold Mining in Grass Valley California - Unavailable since 1940, this publication was originally compiled by the United States Department of Interior. Included are insights into the gold mineralization and other characteristics of this important mining region of Nevada County, California. This volume also includes important insights into the geologic formations, faults and other aspects of economic geology in this California mining district. Of particular interest are the fine details on many hardrock gold mines in the area, including their locations, histories, development and mineralization. Some of the mines featured include the Gold Hill Mine, Massachusetts Hill, Boundary, Peabody, Golden Center, North Star, Omaha, Lone Jack, Homeward Bound, Hartery, Wisconsin, Allison Ranch, Phoenix, Kate Hayes, W.Y.O.D., Empire, Rich Hill, Daisy Hill, Orleans, Sultana, Centennial, Conlin, Ben Franklin, Crown Point and many others. **8.5" X 11", 148 ppgs. Retail Price: $12.99**

Lode Mining in the Alleghany District of Sierra County California - Unavailable since 1913, this publication was originally compiled by the United States Department of Interior. Included are insights into the mineralization and other characteristics of this important mining region of Sierra County. Included are details on the history, production and locations of numerous hardrock gold mines in this famous California area, including the Tightner Mine, Minnie D., Osceola, Eldorado, Twenty One, Sherman, Kenton, Oriental, Rainbow, Plumbago, Irelan, Gold Canyon, North Fork, Federal, Kate Hardy and others. This volume also includes important insights into the fossil record, geologic formations, faults and other aspects of economic geology in this California mining district. **8.5" X 11", 48 ppgs. Retail Price: $7.99**

Six Months In The Gold Mines During The California Gold Rush - Unavailable since 1850, this important work is a first hand account of one "49'ers" personal experience during the great California Gold Rush, shedding important light on one of the most exciting periods in the history of not only California, but also the world. Compiled from journals written between 1847 and 1849 by E. Gould Buffum, a native of New York, "Six Months In The Gold Mines During The California Gold Rush" offers a rare look into the day to day lives of the people who came to California to work in her gold mines when the state was still a great frontier. **8.5" X 11", 290 ppgs. Retail Price: $19.99**

<u>Quartz Mines of the Grass Valley Mining District of California</u> - Unavailable since 1867, this important publication has not been available since those days. This rare publication offers a short dissertation on the early hardrock mines in this important mining district in the California Mother Lode region between the 1850's and 1860's. Also included are hard to find details on the mineralization and locations of these mines, as well as how they were operated in those day. **8.5" X 11", 44 ppgs. Retail Price: $8.99**

<u>Gold Rush on the Feather River</u> - First published in 1924, this short publication by G.C. Mansfield sheds important light on the early history of gold mining on the Feather River. Included are rare insights into the first decade of gold mining and the early mining camps of the Feather River during the 1850's. 64 ppgs., 9.99

<u>The Bodie Mining District of California</u> - First published in 1986, it has been unavailable since those days and sheds important light on this famous mining area. Included are the history, characteristics and locations of numerous old mines around the ghost town of Bodie. 64 ppgs, 8.99

<u>Geology and Mineral Resources of the Gasquet Quadrangle of California-Oregon</u> - First published in 1953, it has been unavailable for over a century and sheds important light on the geological features and mineral resources of this portion of Northern California and Southern Oregon. 80 ppgs, 9.99

Alaska Mining Books

<u>Ore Deposits of the Willow Creek Mining District, Alaska</u> - Unavailable since 1954, this hard to find publication includes valuable insights into the Willow Creek Mining District near Hatcher Pass in Alaska. The publication includes insights into the history, geology and locations of the well known mines in the area, including the Gold Cord, Independence, Fern, Mabel, Lonesome, Snowbird, Schroff-O'Neil, High Grade, Marion Twin, Thorpe, Webfoot, Kelly-Willow, Lane, Holland and others. **8.5" X 11", 96 ppgs. Retail Price: $9.99**

<u>The Juneau Gold Belt of Alaska</u> - Unavailable since 1906, this hard to find publication includes valuable insights into the gold mines around Juneau, Alaska. The publication includes important details into the history, geology and locations of the well known gold mines and prospects in the area, including those around Windham Bay, Holkham Bay, Port Snettisham, on Grindstone and Rhine Creeks, Gold Creek, Douglas Island, Salmon Creek, Lemon Creek, Nugget Creek, from the Mendenhall River to Berners Bay, McGinnis Creek, Montana Creek, Peterson Creek, Windfall Creek, the Eagle River, Yankee Basin, Yankee Curve, Kowee Creek and elsewhere. Not only are gold placer mines included, but also hardrock gold mines. **8.5" X 11", 224 ppgs. Retail Price: $19.99**

<u>Mining in the Jumbo Basin of Alaska</u> - Unavailable since 1953, this hard to find publication includes valuable insights into the mines and geology of the Jumbo Basin. The publication includes important details into the history, geology and locations of the well known gold mines and prospects in the famous Jumbo Basin Mining Region of Alaska. 72 ppgs, 9.99

<u>The Rampart Placer Gold Region of Alaska</u> - Unavailable since 1906, this hard to find publication includes valuable insights into the placer gold mines of the Rampart Mining Region. The publication includes important details into the history, geology and locations of the well known gold mines and prospects in the famous Rampart Mining Region of Alaska. 78 ppgs, 10.99

Arizona Mining Books

<u>Mines and Mining in Northern Yuma County Arizona</u> - Originally published in 1911, this important publication on Arizona Mining has not been available for over a hundred years. Included are rare insights into the gold, silver, copper and quicksilver mines of Yuma County, Arizona together with hard to find maps and photographs. Some of the mines and mining districts featured include the Planet Copper Mine, Mineral Hill, the Clara Consolidated Mine, Viati Mine, Copper Basin prospect, Bowman Mine, Quartz King, Billy Mack, Carnation, the Wardwell and Osbourne, Valensuella Copper, the Mariquita, Colonial Mine, the French American, the New York-Plomosa, Guadalupe, Lead Camp, Mudersbach Copper Camp, Yellow Bird, the Arizona Northern (Salome Strike), Bonanza (Harqua Hala), Golden Eagle, Hercules, Socorro and others. **8.5" X 11", 144 ppgs. Retail Price: $11.99**

<u>The Aravaipa and Stanley Mining Districts of Graham County Arizona</u> - Originally published in 1925, this important publication on Arizona Mining has not been available for nearly ninety years. Included are rare insights into the gold and silver mines of these two important mining districts, together with hard to find maps. **8.5" X 11", 140 ppgs. Retail Price: $11.99**

Gold in the Gold Basin and Lost Basin Mining Districts of Mohave County, Arizona - This volume contains rare insights into the geology and gold mineralization of the Gold Basin and Lost Basin Mining Districts of Mohave County, Arizona that will be of benefit to miners and prospectors. Also included is a significant body of information on the gold mines and prospects of this portion of Arizona. This volume is lavishly illustrated with rare photos and mining maps. **8.5" X 11", 188 ppgs. Retail Price: $19.99**

Mines of the Jerome and Bradshaw Mountains of Arizona - This important publication on Arizona Mining has not been available for ninety years. This volume contains rare insights into the geology and ore deposits of the Jerome and Bradshaw Mountains of Arizona that will be of benefit to miners and prospectors who work those areas. Included is a significant body of information on the mines and prospects of the Verde, Black Hills, Cherry Creek, Prescott, Walker, Groom Creek, Hassayampa, Bigbug, Turkey Creek, Agua Fria, Black Canyon, Peck, Tiger, Pine Grove, Bradshaw, Tintop, Humbug and Castle Creek Mining Districts. This volume is lavishly illustrated with rare photos and mining maps. **8.5" X 11", 218 ppgs. Retail Price: $19.99**

The Ajo Mining District of Pima County Arizona - This important publication on Arizona Mining has not been available for nearly seventy years. This volume contains rare insights into the geology and mineralization of the Ajo Mining District in Pima County, Arizona and in particular the famous New Cornelia Mine. **8.5" X 11", 126 ppgs. Retail Price: $11.99**

Mining in the Santa Rita and Patagonia Mountains of Arizona - Originally published in 1915, this important publication on Arizona Mining has not been available for nearly a century. Included are rare insights into hundreds of gold, silver, copper and other mines in this famous Arizona mining area. Details include the locations, geology, history, production and other facts of the mines of this region. **8.5" X 11", 394 ppgs. Retail Price: $24.99**

Mining in the Bisbee Quadrangle of Arizona - Originally published in 1906, this important publication on Arizona Mining has not been available for nearly a century. Included are rare insights into hundreds of gold, silver, copper and other mines in this famous Arizona mining area. Details include the locations, geology, history, production and other facts of the mines of this important mining region. **8.5" X 11", 188 ppgs. Retail Price: $14.99**

Placer Gold Mining in Arizona - Unavailable since 1922, this hard to find publication includes valuable insights into the placer gold mines of the Arizona. Originally released as "Placer Gold of Arizona", despite its small size, this publication includes important details into the history, geology and locations of the well known placer gold mines and prospects in the State of Arizona. **48 ppgs, 8.99**

Gold and Copper Mining near Payson, Arizona - Written in 1915, this hard to find publication includes valuable insights into the gold and copper mining industry of Arizona. Highlighted here are the gold and copper mines near Payson, Arizona. **68 ppgs, 8.99**

Lode Gold Mining in Arizona - Unavailable since 1934, this hard to find publication, originally released as "Arizona Lode Gold Mines and Gold Mining" includes valuable insights into the gold mining industry of Arizona. Included are valuable insights into over 150 hardrock gold mines in over 30 different mining districts in Arizona. **278 ppgs, 21.99**

Mining in the Dragoon Quadrangle of Cochise County, Arizona - Unavailable since 1964, this hard to find publication includes valuable insights into the mines of the Dragoon Quadrangle Mining Region. The publication includes important details into the history, geology and locations of the well known mines and prospects in this famous mining region of Arizona. **224 ppgs., 19.99**

Directory of Operating Mines in Arizona in 1915 - Unavailable since 1916, this hard to find publication includes valuable insights into the mines of Arizona. This small publication includes a complete list of the mines that were operating in the State of Arizona during 1915 and includes details such as general location, owners and some basic facts about each mining operation. **52 ppgs. 8.99**

Arizona Ore Deposits - Unavailable since 1938, this hard to find publication includes valuable insights into some ore deposits of Arizona. Included are valuable insights into the formation and characteristics of valuable ore deposits in the Jerome, Miami, Inspiration, Clifton, Morenci, Ray, Ajo, Eureka, Tombstone and Magma mining districts. Included are details into some of the major gold, silver and copper mines of these important Arizona mining areas. **160 ppgs, 14.99**

Montana Mining Books

A History of Butte Montana: The World's Greatest Mining Camp - First published in 1900 by H.C. Freeman, this important publication sheds a bright light on one of the most important mining areas in the history of The West. Together with his insights, as well as rare photographs of the periods, Harry Freeman describes Butte and its vicinity from its early beginnings, right up to its flush years when copper flowed from its mines like a river. At the time of publication, Butte, Montana was known worldwide as "The Richest Mining Spot On Earth" and produced not only vast amounts of copper, but also silver, gold and other metals from its mines. Freeman illustrates, with great detail, the most important mines in the vicinity of Butte, providing rare details on their owners, their history and most importantly, how the mines operated and how their treasures were extracted. Of particular interest are the dozens of rare photographs that depict mines such as the famous Anaconda, the Silver Bow, the Smoke House, Moose, Paulin, Buffalo, Little Minah, the Mountain Consolidated, West Greyrock, Cora, the Green Mountain, Diamond, Bell, Parnell, the Neversweat, Nipper, Original and many others. **8.5" X 11", 142 ppgs. Retail Price: $12.99**

The Butte Mining District of Montana - This important publication on Montana Mining has not been available for over a century. Included are rare insights into the gold, copper and silver mines of Butte, Montana together with hard to find maps and photographs. Some of the topics include the early history of gold, silver and copper mining in the Butte area, insight into the geology of its mining areas, the local distribution of gold, silver and copper ores, as well their composition and how to identify them. Also included are detailed facts about the mines in the Butte Mining District, including the famous Anaconda Mine, Gagnon, Parrot, Blue Vein, Moscow, Poulin, Stella, Buffalo, Green Mountain, Wake Up Jim, the Diamond-Bell Group, Mountain Consolidated, East Greyrock, West Greyrock, Snowball, Corra, Speculator, Adirondack, Miners Union, the Jessie-Edith May Group, Otisco, Iduna, Colorado, Lizzie, Cambers, Anderson, Hesperus, Preferencia and dozens of others. **8.5" X 11", 298 ppgs. Retail Price: $24.99**

Mines of the Helena Mining Region of Montana - This important publication on Montana Mining has not been available for over a century. Included are rare insights into the gold, copper and silver mines of the vicinity of Helena, Montana, including the Marysville Mining District, Elliston Mining District, Rimini Mining District, Helena Mining District, Clancy Mining District, Wickes Mining District, Boulder and Basin Mining Districts and the Elkhorn Mining District. Some of the topics include the early history of gold, silver and copper mining in the Helena area, insight into the geology of its mining areas, the local distribution of gold, silver and copper ores, as well their composition and how to identify them. Also included are detailed facts, history, geology and locations of over one hundred gold, silver and copper mines in the area . **8.5" X 11", 162 ppgs, Retail Price: $14.99**

Mines and Geology of the Garnet Range of Montana - This important publication on Montana Mining has not been available for over a century. Included are rare insights into the gold, copper and silver mines of the vicinity of this important mining area of Montana. Some of the topics include the early history of gold, silver and copper mining in the Garnet Mountains, insight into the geology of its mining areas, the local distribution of gold, silver and copper ores, as well their composition and how to identify them. Also included are detailed facts, history, geology and locations of numerous gold, silver and copper mines in the area . **8.5" X 11", 100 ppgs, Retail Price: $11.99**

Mines and Geology of the Philipsburg Quadrangle of Montana - This important publication on Montana Mining has not been available for over a century. Included are rare insights into the gold, copper and silver mines of the vicinity of this important mining area of Montana. Some of the topics include the early history of gold, silver and copper mining in the Philipsburg Quadrangle, insight into the geology of its mining areas, the local distribution of gold, silver and copper ores, as well their composition and how to identify them. Also included are detailed facts, history, geology and locations of over one hundred gold, silver and copper mines in the area **8.5" X 11", 290 ppgs, Retail Price: $24.99**

Geology of the Marysville Mining District of Montana - Included are rare insights into the mining geology of the Marysville Mining District. Some of the topics include the early history of gold, silver and copper mining in the area, insight into the geology of its mining areas, the local distribution of gold, silver and copper ores, as well their composition and how to identify them. Also included are detailed facts, history, geology and locations of gold, silver and copper mines in the area **8.5" X 11", 198 ppgs, Retail Price: $19.99**

The Geology and Mines of Northern Idaho and North Western Montana- See listing under Idaho.

The History of Gold Dredging in Montana - Unavailable since 1916, this important publication was originally published by the Us Bureau of Mines and has been unavailable for a century. A century and more ago, giant dredging machines dug in Montana's rivers and creeks in search of illusive golden riches. First appearing in California in the 1850's, gold dredges finally reached their peak of development in Siberia and New Zealand before becoming popular again in the United States. This book offers a unique historical perspective on the gold dredges that once operated in Montana. This book on Montana mining history is lavishly illustrated with dozens of rare historic photos gold dredges that once operated in Montana, as well as hard to locate plans on how these dredges were designed. 120 ppgs., 11.99

Nevada Mining Books

The Bull Frog Mining District of Nevada - Unavailable since 1910, this publication was originally compiled by the United States Department of Interior. This volume also includes important insights into the geologic formations, faults and other aspects of economic geology in this Nevada mining district. Of particular interest are the fine details on many mines in the area, including their locations, histories, development and mineralization. Some of the mines featured include the National Bank Mine, Providence, Gibraltor, Tramps, Denver, Original Bullfrog, Gold Bar, Mayflower, Homestake-King and other mines and prospects. **8.5" X 11", 152 ppgs, Retail Price: $14.99**

History of the Comstock Lode - Unavailable since 1876, this publication was originally released by John Wiley & Sons. This volume also includes important insights into the famous Comstock Lode of Nevada that represented the first major silver discovery in the United States. During its spectacular run, the Comstock produced over 192 million ounces of silver and 8.2 million ounces of gold. Not only did the Comstock result in one of the largest mining rushes in history and yield immense fortunes for its owners, but it made important contributions to the development of the State of Nevada, as well as neighboring California. Included here are important details on not only the early development and history of the Comstock, but also rare early insight into its mines, ore and its geology.**8.5" X 11", 244 ppgs, Retail Price: $19.99**

The Pioche Mining District of Nevada - First published in 1932, it has been unavailable for over a century and sheds important light on the mining history of Nevada. Some of the topics include the history of mining in this district, as well as the characteristics of its mineral and ore deposits. Also included are insights into the history, production, characteristics and locations of numerous mines in the area. Some of the mines include the Combined Metals, Pioche, Ely Valley, No. 10, Poorman, Wide Awake, Alps, Prince, Virginia Louise, Half Moon, Abe Lincoln, Fairview, Bristol Silver, National, Vesuvius, Inman, Tempest, Hillside, Jackrabbit, Lucky Star, Fortuna, Mendha, Manhattan, Hamburg, Comet, Lyndon and others. 108 ppgs 10.99

The Yerington Mining District of Nevada - First published in 1932, it has been unavailable for over a century and sheds important light on the mining history of Nevada. Some of the topics include the history of mining in this district, as well as the characteristics of its mineral and ore deposits. Also included are insights into the history, production, characteristics and locations of numerous mines in the area. Some of the mines include the Bluestone, Mason Valley, Malachite, McConnell, Greenwood, Western Nevada, Ludwig, Douglas Hill, Casting Copper, Montana-Yerington, Empire, Jim Beatty, Terry and McFarland, Blue Jay and others. 92 ppgs, 10.99

The Genesis of the Ores of Tonopah Nevada - Unavailable since 1918, this hard to find publication includes valuable insights into the gold mines around Tonopah, Nevada. The publication includes important details into the geology of mines in the Tonopah Mining District of Nevada. 90 ppgs, 10.99

Mining Camps of Elko, Lander and Eureka Counties Nevada - Unavailable since 1910, this hard to find publication includes valuable insights into the mining camps of Elko, Lander and Eureka Counties, Nevada. The publication includes important details into the history of mines and mining in these three Nevada counties. 154 ppgs, 12.99

Ore Deposits of the Bullfrog Quadrangle - Unavailable since 1964 and released as "Geology of Bullfrog Quadrangle and Ore Deposits Related to Bullfrog Hills Caldera, Nye County, Nevada and Inyo County, California". The publication includes important details into the geology of mines in the Bullfrog Quadrangle of Nye County, Nevada and Inyo County, California. 52 ppgs, 9.99

Mining in Eureka County Nevada - Unavailable since 1879, this hard to find publication includes valuable insights into the early mining history off Eureka County, Nevada. The publication includes important details into the early history of the mines of Eureka County, as well as their development, production and how their ores were treated. Also included are details on the 1872 Mining Act, as well as the local rules, regulations and customs of the miners in Eureka County.134 ppgs, 12.99

Colorado Mining Books

Ores of The Leadville Mining District - Unavailable since 1926, this publication was originally compiled by the United States Department of Interior. This volume also includes important insights into the ores and mineralization of the Leadville Mining District in Colorado. Topics include historic ore prospecting methods, local geology, insights into ore veins and stockworks, the local trend and distribution of ore channels, reverse faults, shattered rock above replacement ore bodies, mineral enrichment in oxidized and sulphide zones and more. **8.5" X 11", 66 ppgs, Retail Price: $8.99**

Mining in Colorado - Unavailable since 1926, this publication was originally compiled by the United States Department of Interior. This volume also includes important insights into the mining history of Colorado from its early beginnings in the 1850's right up to the mid 1920's. Not only is Colorado's gold mining heritage included, but also its silver, copper, lead and zinc mining industry. Each mining area is treated separately, detailing the development of Colorado's mines on a county by county basis. **8.5" X 11", 284 ppgs, Retail Price: $19.99**

Gold Mining in Gilpin County Colorado - Unavailable since 1876, this publication was originally compiled by the Register Steam Printing House of Central City, Colorado. A rare glimpse at the gold mining history and early mines of Gilpin County, Colorado from their first discovery in the 1850's up to the "flush years" of the mid 1870's. Of particular interest is the history of the discovery of gold in Gilpin County and details about the men who made those first strikes. Special focus is given to the early gold mines and first mining districts of the area, many of which are not detailed in other books on Colorado's gold mining history. **8.5" X 11", 156 ppgs, Retail Price: $12.99**

Mining in the Gold Brick Mining District of Colorado - Important insights into the history of the Gold Brick Mining District, as well as its local geography and economic geology. Also included are the histories and locations of historic mines in this important Colorado Mining District, including the Cortland, Carter, Raymond, Gold Links, Sacramento, Bassick, Sandy Hook, Chronicle, Grand Prize, Chloride, Granite Mountain, Lucille, Gray Mountain, Hilltop, Maggie Mitchell, Silver Islet, Revenue, Roosevelt, Carbonate King and others. In addition to hardrock mining, are also included are details on gold placer mining in this portion of Colorado. **8.5" X 11", 140 ppgs, Retail Price: $12.99**

Ore Deposits of the London Fault of Colorado - First published in 1941, it has been unavailable since those days and sheds important light on the mines and mineral deposits of the London Fault in Central Colorado's Alma Mining District. This publication sheds important light on the gold veins and lead-silver deposits of the Alma Mining District. Included are geologic details on the London Mine, American Mine, Havigorst Tunnel, Ophir Mine, Mosher Tunnel, London-Butte Mine, Venture Shaft, Hard-To-Beat Mine, Oliver Twist Tunnel, Sacramento Mine, Mudsill Mine, Sherwood Mine, Wagner, Barcoe Tunnel and other mines in this important mining region. 110 ppgs., 10.99

The Mines of Colorado - First published in 1867, it has been unavailable since those days and sheds important light on Colorado's early mining history. Written shortly after the events took place, this publication sheds important light on the Pike's Peak Gold Rush, the discovery of gold on Ralston Creek and Dry Creek in the 1850's, as well as details on the first wave of miners into Colorado and their trials and tribulations as they crossed the Great Plains. Also included are details on early discoveries of lode gold in the mountainous regions of Colorado, details on the early mines hardrock and placer mines, and much more. It is a veritable treasure trove on Colorado's early mining history and will be of great importance to anyone who is interested in the mining of gold or other minerals in Colorado, as well as those interested in the history of the state. 478 ppgs., 29.99

The La Plata Mining District of Colorado - Originally titled "Geology and Ore Deposits in the Vicinity of the La Plata District of Colorado" and first published in 1949, it has been unavailable since those days and sheds important light on the mines and mineral deposits of the La Plata Mining District of Colorado. 214 ppgs., 19.99

Washington Mining Books

The Republic Mining District of Washington - Unavailable since 1910, this important publication was originally published by the Washington Geologic Survey and has been unavailable for a century. Topics include the geology, rock formations and the formation of ore deposits in this important mining area of Washington State. Also included are hard to find details on the geology, history and locations of dozens of mines in the area. Some of the mines featured include the New Republic Mine, Ben Hur, Morning Glory, the South Republic Mine, Quilp, Surprise, Black Tail, Lone Pine, San Poil, Mountain Lion, Tom Thumb, Elcaliph and many others. **8.5" X 11", 94 ppgs, Retail Price: $10.99**

The Myers Creek and Nighthawk Mining Districts of Washington - Unavailable since 1911, this important publication was originally published by the Washington Geologic Survey and has been unavailable for a century. Topics include the geology, rock formations and the formation of ore deposits in these important mining areas of Washington State. Also included are hard to find details on the geology, history and locations of dozens of mines in the area. Some of the mines featured include the Grant Mine, Monterey, Nip and Tuck, Myers Creek, Number Nine, Neutral, Rainbow, Aztec, Crystal Butte, Apex, Butcher Boy, Molson, Mad River, Olentangy, Delate, Kelsey, Golden Chariot, Okanogan, Ohio, Forty-Ninth Parallel, Nighthawk, Favorite, Little Chopaka, Summit, Number One, California, Peerless, Caaba, Prize Group, Ruby, Mountain Sheep, Golden Zone, Rich Bar, Similkameen, Kimberly, Triune, Hiawatha, Trinity, Hornsilver, Maquae, Bellevue, Bullfrog, Palmer Lake, Ivanhoe, Copper World and many others. **8.5" X 11", 136 ppgs, Retail Price: $12.99**

The Blewett Mining District of Washington - Unavailable since 1911, this important publication was originally published by the Washington Geologic Survey and has been unavailable for a century. Topics include the geology, rock formations and the formation of ore deposits in this important mining area of Washington State. Also included are hard to find details on the geology, history and locations of dozens of mines in the area. Some of the mines featured include the Washington Meteor, Alta Vista, Pole Pick, Blinn, North Star, Golden Eagle, Tip Top, Wilder, Golden Guinea, Lucky Queen, Blue Bell, Prospect, Homestake, Lone Rock, Johnson, and others. **8.5" X 11", 134 ppgs, Retail Price: $12.99**

Silver Mining In Washington - Unavailable since 1955, this important publication was originally published by the Washington Geologic Survey. Featured are the hard to find locations and details pertaining to Washington's silver mines. **8.5" X 11", 180 ppgs, Retail Price: $15.99**

The Mines of Snohomish County Washington - Unavailable since 1942, this important publication was originally published by the Washington Geologic Survey and has been unavailable for seventy years. Featured are details on a large number of gold, silver, copper, lead and other metallic mineral mines. Included are the locations of each historic mine, along with information on the commodity produced. **8.5" X 11", 98 ppgs, Retail Price: $10.99**

The Mines of Chelan County Washington - Unavailable since 1943, this important publication was originally published by the Washington Geologic Survey and has been unavailable for seventy years. Featured are details on a large number of gold, silver, copper, lead and other metallic mineral mines. Included are the locations of each historic mine, along with information on the commodity. **8.5" X 11", 88 ppgs, Retail Price: $9.99**

Metal Mines of Washington - Unavailable since 1921, this important publication was originally published by the Washington Geologic Survey and has been unavailable for nearly ninety years. Widely considered a masterpiece on the Washington Mining Industry, "Metal Mines of Washington" sheds light on the important details of Washington's early mining years. Featured are details on hundreds of gold, silver, copper, lead and other metallic mineral mines. Included are hard to find details on the mineral resources of this state, as well as the locations of historic mines. Lavishly illustrated with maps and historic photos and complete with a glossary to explain any technical terms found in the text, this is one of the most important works on mining in the State of Washington. No prospector or miner should be without it if they are interested in mining in Washington. **8.5" X 11", 396 ppgs, Retail Price: $24.99**

Gem Stones In Washington - Unavailable since 1949, this important publication was originally published by the Washington Geologic Survey and has been unavailable since first published. Included are details on where to find naturally occurring gem stones in the State of Washington, including quartz crystal, amethyst, smoky quartz, milky quartz, agates, bloodstone, carnelian, chert, flint, jasper, onyx, petrified wood, opal, fire opal, hyalite and others. **8.5" X 11", 54 ppgs, Retail Price: $8.99**

The Covada Mining District of Washington - Unavailable since 1913, this important publication was originally published by the Washington Geologic Survey and has been unavailable for a century. Topics include the geology, rock formations and the formation of ore deposits in this important mining area of Washington State. Also included are hard to find details on the geology, history and locations of dozens of mines in the area. Some of the mines featured include the Admiral, Advance, Algonkian, Big Bug, Big Chief, Big Joker, Black Hawk, Black Tail, Black Thorn, Captain, Cherokee Strip, Colorado, Dan Patch, Dead Shot, Etta, Good Ore, Greasy Run, Great Scott, Idora, IXL, Jay Bird, Kentucky Bell, King Solomon, Laurel, Laura S, Little Jay, Meteor, Neglected, Northern Light, Old Nell, Plymouth Rock, Polaris, Quandary, Reserve, Shoo Fly, Silver Plume, Three Pines, Vernie, White Rose and dozens of others. **8.5" X 11", 114 ppgs, Retail Price: $10.99**

The Index Mining District of Washington - Unavailable since 1912, this important publication was originally published by the Washington Geologic Survey and has been unavailable for a century. Topics include the geology, rock formations and the formation of ore deposits in this important mining area of Washington State. Also included are hard to find details on the geology, history and locations of dozens of mines in the area. Some of the mines featured include the Sunset, Non-Pareil, Ethel Consolidated, Kittaning, Merchant, Homestead, Co-operative, Lost Creek, Uncle Sam, Calumet, Florence-Rae, Bitter Creek, Index Peacock, Gunn Peak, Helena, North Star, Buckeye. Copper Bell, Red Cross and others. **8.5″ X 11″, 114 ppgs, Retail Price: $11.99**

Mining & Mineral Resources of Stevens County Washington - Unavailable since 1920, this important publication was originally published by the Washington Geologic Survey and has been unavailable for a century. Topics include the geology, rock formations and the formation of ore deposits in these important mining areas of Washington State. Also included are hard to find details on the geology, history and locations of hundreds of mines in the area. **8.5″ X 11″, 372 ppgs, Retail Price: $24.99**

The Mines and Geology of the Loomis Quadrangle Okanogan County, Washington - Unavailable since 1972, this important publication was originally published by the Washington Geologic Survey and has been unavailable for a century. Topics include the geology, rock formations and the formation of ore deposits in this important mining area of Washington State. Also included are hard to find details on the geology, history and locations of dozens of gold, copper, silver and other mines in the area. **8.5″ X 11″, 150 ppgs, Retail Price: $12.99**

The Conconully Mining District of Okanogan County Washington - Unavailable since 1973, this important publication was originally published by the Washington Geologic Survey and has been unavailable for a century. Topics include the geology, rock formations and the formation of ore deposits in this important mining area of Washington State, which also includes Salmon Creek, Blue Lake and Galena. Also included are hard to find details on the geology, mining history and locations of dozens of mines in the area. Some of the mines include Arlington, Fourth of July, Sonny Boy, First Thought, Last Chance, War Eagle-Peacock, Wheeler, Mohawk, Lone Star, Woo Loo Moo Loo, Keystone, Hughes, Plant-Callahan, Johnny Boy, Leuena, Gubser, John Arthur, Tough Nut, Homestake, Key and many others **8.5″ X 11″, 68 ppgs, Retail Price: $8.99**

Wyoming Mining Books

Mining in the Laramie Basin of Wyoming - Unavailable since 1909, this publication was originally compiled by the United States Department of Interior. Also included are insights into the mineralization and other characteristics of this important mining region, especially in regards to coal, limestone, gypsum, bentonite clay, cement, sand, clay and copper. **8.5″ X 11″, 104 ppgs, Retail Price: $11.99**

New Mexico Mining Books

The Mogollon Mining District of New Mexico - Unavailable since 1927, this important publication was originally published by the US Department of Interior and has been unavailable for 80 years. Topics include the geology, rock formations and the formation of ore deposits in this important mining area in New Mexico. Of particular focus is information on the history and production of the ore deposits in this area, their form and structure, vein filling, their paragenesis, origins and ore shoots, as well as oxidation and supergene enrichment. Also included are hard to find details, including the descriptions and locations of numerous gold, silver and other types of mines, including the Eureka, Pacific, South Alpine, Great Western, Enterprise, Buffalo, Mountain View, Floride, Gold Dust, Last Chance, Deadwood, Confidence, Maud S., Deep Down, Little Fanney, Trilby, Johnson, Alberta, Comet, Golden Eagle, Cooney, Queen, the Iron Crown, Eberle, Clifton, Andrew Jackson mine, Mascot and others. **8.5″ X 11″, 144 ppgs, Retail Price: $12.99**

The Percha Mining District of Kingston New Mexico - Unavailable since 1883, this important publication was originally published by the Kingston Tribune and has been unavailable for over one hundred and thirty five years. Having been written during the earliest years of gold and silver mining in the Percha Mining District, unlike other books on the subject, this work offers the unique perspective of having actually been written while the early mining history of this area was still being made. In fact, the work was written so early in the development of this area that many of the notable mines in the Percha District were less than a few years old and were still being operated by their original discoverers with the same enthusiasm as when they were first located. Included are hard to find details on the very earliest gold and silver mines of this important mining district near Kingston in Sierra County, New Mexico. **8.5″ X 11″, 68 ppgs, Retail Price: $9.99**

East Coast Mining Books

<u>The Gold Fields of the Southern Appalachians</u> - Unavailable since 1895, this important publication was originally published by the US Department of Interior and has been unavailable for nearly 120 years. Topics include the geology, rock formations and the formation of ore deposits in this important mining area of the American South. Of particular focus is information on the history and statistics of the ore deposits in this area, their form and structure and veins. Also included are details on the placer gold deposits of the region. The gold fields of the Georgian Belt, Carolinian Belt and the South Mountain Mining District of North Carolina are all treated in descriptive detail. Included are hard to find details, including the descriptions and locations of numerous gold mines in Georgia, North Carolina and elsewhere in the American South. Also included are details on the gold belts of the British Maritime Provinces and the Green Mountains. **8.5" X 11", 104 ppgs, Retail Price: $9.99**

Gold Rush Tales Series

<u>**Millions in Siskiyou County Gold**</u> - In this first volume of the "Gold Rush Tales" series, leading mining historian and editor Kerby Jackson, introduces us to the story of how millions of dollars worth of gold was discovered in Siskiyou County during the California Gold Rush. Lavishly illustrated with photos from the 19th Century, this hard to find information was first published in 1897 and sheds important light onto the gold rush era in Siskiyou County, California and the experiences of the men who dug for the gold and actually found it. **8.5" X 11", 82 ppgs, Retail Price: $9.99**

<u>**The California Rand in the Days of '49**</u> - In this second volume of the "Gold Rush Tales" series, leading mining historian and editor Kerby Jackson, introduces us to four tales from the California Gold Rush. Lavishly illustrated with photos from the 19th Century, this hard to find information was first published in 1890's and includes the stories of "California's Rand", details about Chinese miners, how one early miner named Baker struck it rich and also the story of Alphonzo Bowers, who invented the first hydraulic gold dredge. **8.5" X 11", 54 ppgs, Retail Price: $9.99**

More Mining Books

<u>**Prospecting and Developing A Small Mine**</u> - Topics covered include the classification of varying ores, how to take a proper ore sample, the proper reduction of ore samples, alluvial sampling, how to understand geology as it is applied to prospecting and mining, prospecting procedures, methods of ore treatment, the application of drilling and blasting in a small mine and other topics that the small scale miner will find of benefit. **8.5" X 11", 112 ppgs, Retail Price: $11.99**

<u>**Timbering For Small Underground Mines**</u> - Topics covered include the selection of caps and posts, the treatment of mine timbers, how to install mine timbers, repairing damaged timbers, use of drift supports, headboards, squeeze sets, ore chute construction, mine cribbing, square set timbering methods, the use of steel and concrete sets and other topics that the small underground miner will find of benefit. This volume also includes twenty eight illustrations depicting the proper construction of mine timbering and support systems that greatly enhance the practical usability of the information contained in this small book. **8.5" X 11", 88 ppgs. Retail Price: $10.99**

<u>**Timbering and Mining**</u> - A classic mining publication on Hard Rock Mining by W.H. Storms. Unavailable since 1909, this rare publication provides an in depth look at American methods of underground mine timbering and mining methods. Topics include the selection and preservation of mine timbers, drifting and drift sets, driving in running ground, structural steel in mine workings, timbering drifts in gravel mines, timbering methods for driving shafts, positioning drill holes in shafts, timbering stations at shafts, drainage, mining large ore bodies by means of open cuts or by the "Glory Hole" system, stoping out ore in flat or low lying veins, use of the "Caving System", stoping in swelling ground, how to stope out large ore bodies, Square Set timbering on the Comstock and its modifications by California miners, the construction of ore chutes, stoping ore bodies by use of the "Block System", how to work dangerous ground, information on the "Delprat System" of stoping without mine timbers, construction and use of headframes and much more. This volume provides a reference into not only practical methods of mining and timbering that may be employed in narrow vein mining by small miners today, but also rare insights into how mines were being worked at the turn of the 19th Century. **8.5" X 11", 288 ppgs. Retail Price: $24.99**

A Study of Ore Deposits For The Practical Miner - Mining historian Kerby Jackson introduces us to a classic mining publication on ore deposits by J.P. Wallace. First published in 1908, it has been unavailable for over a century. Included are important insights into the properties of minerals and their identification, on the occurrence and origin of gold, on gold alloys, insights into gold bearing sulfides such as pyrites and arsenopyrites, on gold bearing vanadium, gold and silver tellurides, lead and mercury tellurides, on silver ores, platinum and iridium, mercury ores, copper ores, lead ores, zinc ores, iron ores, chromium ores, manganese ores, nickel ores, tin ores, tungsten ores and others. Also included are facts regarding rock forming minerals, their composition and occurrences, on igneous, sedimentary, metamorphic and intrusive rocks, as well as how they are geologically disturbed by dikes, flows and faults, as well as the effects of these geologic actions and why they are important to the miner. Written specifically with the common miner and prospector in mind, the book will help to unlock the earth's hidden wealth for you and is written in a simple and concise language that anyone can understand. **8.5" X 11", 366 ppgs. Retail Price: $24.99**

Mine Drainage - Unavailable since 1896, this rare publication provides an in depth look at American methods of underground mine drainage and mining pump systems. This volume provides a reference into not only practical methods of mining drainage that may be employed in narrow vein mining by small miners today, but also rare insights into how mines were being worked at the turn of the 19th Century. **8.5" X 11", 218 ppgs. Retail Price: $24.99**

Fire Assaying Gold, Silver and Lead Ores - Unavailable since 1907, this important publication was originally published by the Mining and Scientific Press and was designed to introduce miners and prospectors of gold, silver and lead to the art of fire assaying. Topics include the fire assaying of ores and products containing gold, silver and lead; the sampling and preparation of ore for an assay; care of the assay office, assay furnaces; crucibles and scorifiers; assay balances; metallic ores; scorification assays; cupelling; parting' crucible assays, the roasting of ores and more. This classic provides a time honored method of assaying put forward in a clear, concise and easy to understand language that will make it a benefit to even beginners. **8.5" X 11", 96 ppgs. Retail Price: $11.99**

Methods of Mine Timbering - Originally published in 1896, this important publication on mining engineering has not been available for nearly a century. Included are rare insights into historical methods of timbering structural support that were used in underground metal mines during the California that still have a practical application for the small scale hardrock miner of today. **8.5" X 11", 94 ppgs. Retail Price: $10.99**

The Enrichment of Copper Sulfide Ores - First published in 1913, it has been unavailable for over a century. Topics include the definition and types of ore enrichment, the oxidation of copper ores, the precipitation of metallic sulfides. Also included are the results of dozens of lab experiments pertaining to the enrichment of sulfide ores that will be of interest to the practical hard rock mine operator in his efforts to release the metallic bounty from his mine's ore. **8.5" X 11", 92 ppgs. Retail Price: $9.99**

A Study of Magmatic Sulfide Ores - Unavailable since 1914, this rare publication provides an in depth look at magmatic sulfide ores. Some of the topics included are the definition and classification of magmatic ores, descriptions of some magmatic sulfide ore deposits known at the time of publication including copper and nickel bearing pyrrhotic ore bodies, chalcopyrite-bornite deposits, pyritic deposits, magnetite-ileminite deposits, chromite deposits and magmatic iron ore deposits. Also included are details on how to recognize these types of ore deposits while prospecting for valuable hardrock minerals. **8.5" X 11", 138 ppgs. Retail Price: $11.99**

The Cyanide Process of Gold Recovery - Unavailable since 1894 and released under the name "The Cyanide Process: Its Practical Application and Economical Results", this rare publication provides an in depth look at the early use of cyanide leaching for gold recovery from hardrock mine ores. This volume provides a reference into the early development and use of cyanide leaching to recover gold. **8.5" X 11", 162 ppgs. Retail Price: $14.99**

California Gold Milling Practices - Unavailable since 1895 and released under the name "California Gold Practices", this rare publication provides an in depth look at early methods of milling used to reduce gold ores in California during the late 19th century. This volume provides a reference into the early development and use of milling equipment during the earliest years of the California Gold Rush up to the age of the Industrial Revolution. Much of the information still applies today and will be of use to small scale miners engaging in hardrock mining. **8.5" X 11", 104 ppgs. Retail Price: $10.99**

Leaching Gold and Silver Ores With The Plattner and Kiss Processes - Mining historian Kerby Jackson introduces us to a classic mining publication on the evaluation and examination of mines and prospects by C.H. Aaron. First published in 1881, it has been unavailable for over a century and sheds important light on the leaching of gold and silver ores with the Plattner and Kiss processes. **8.5" X 11", 204 ppgs. Retail Price: $15.99**

The Metallurgy of Lead and the Desilverization of Base Bullion - First published in 1896, it has been unavailable for over a century and sheds important light on the the recovery of silver from lead based ores. Some of the topics include the properties of lead and some of its compounds, lead ores such as galenite, anglesite, cerussite and others, the distribution of lead ores throughout the United States and the sampling and assaying of lead ores. Also covered is the metallurgical treatment of lead ores, as well as the desilverization of lead by the Pattinson Process and the Parkes Process. Hofman's text has long been considered one of the most important early works on the recovery of silver from lead based ores. 8.5" X 11", 452 ppgs. Retail Price: $29.99

Ore Sampling For Small Scale Miners - First published in 1916, it has been unavailable for over a century and sheds important light on historic methods of ore sampling in hardrock mines. Topics include how to take correct ore samples and the conditions that affect sampling, such as their subdivision and uniformity. Particular detail is given to methods of hand sampling ore bodies by grab sample, pipe sample and coning, as well as sampling by mechanical methods. Also given are insights into the screening, drying and grinding processes to achieve the most consistent sample results and much more. 8.5" X 11", 124 ppgs. Retail Price: $12.99

The Extraction of Silver, Copper and Tin from Ores - First published in 1896, it has been unavailable for over a century and sheds important light on how historic miners recovered silver, copper and tin from their mining operations. The book is split into three sections, including a discussion on the Lixiviation of Silver Ores, the mining and treatment of copper ores as practiced at Tharsis, Spain and the smelting of tin as it was practiced by metallurgists at Pulo Brani, Singapore. Also included is an overview and analysis of these historic metal recovery methods that will be of benefit to those interested in the extraction of silver, copper and tin from small mines. 8.5" X 11", 118 ppgs. Retail Price: $14.99

The Roasting of Gold and Silver Ores - First published in 1880, it has been unavailable for over a century and sheds important light on how historic miners recovered gold and silver rom their mining operations. Topics include details on the most important silver and free milling gold ores, methods of desulphurization of ores, methods of deoxidation, the chlorination of ores, methods and details on roasting gold and silver ores, notes on furnaces and more. Also included are details on numerous methods of gold and silver recovery, including the Ottokar Hofman's Process, the Patera Process, Kiss Process, Augustin Process, Ziervogel Process and others. 8.5" X 11", 178 ppgs. Retail Price: $19.99

The Examination of Mines and Prospects - First published in 1912, it has been unavailable for over a century and sheds important light on how to examine and evaluate hardrock mines, prospects and lode mining claims. Sections include Mining Examinations, Structural Geology, Structural Features of Ore Deposits, Primary Ores and their Distribution, Types of Primary Ore Deposits, Primary Ore Shoots, The Primary Alteration of Wall Rocks, Alterations by Surface Agencies, Residual Ores and their Distribution, Secondary Ores and Ore Shoots and Vein Outcrops. This hard to find information is a must for those who are interested in owning a mine or who already own a lode mining claim and wish to succeed at quartz mining. 8.5" X 11", 250 ppgs. Retail Price: $19.99

Garnets: Their Mining, Milling and Utilization - First published in 1925, it has been unavailable since those days and sheds important light on the mining, milling and utilization of garnets. Included are details on the characteristics of garnets, where they are found and how they were mined. 78 ppgs, 10.99

Gemstones and Precious Stones of North America - Leading mining historian Kerby Jackson introduces us to a classic mining publication on the gems and precious stones of the United States, Canada and mexico. First published in 1890, it has been unavailable since those days and sheds important light on the gems and precious stones that may be found in North America. Included are chapters on diamonds, corundum, sapphire, ruby, topaz, emerald, disapore, spinel, turquoise, tourmaline, garnets, beyrl, peridot, zircon, quartz crystals, feldspars, pearls and many others. Included are details on where these gems and precious stones may be found throughout North America, as well as their characteristics. 360 ppgs, 24.99

Mining Camps and Mining Districts - First released in 1885 by Charles Howard Shinn under the title "Mining Camps: A Study in American Frontier Government", this publication offers a unique look at how early gold miners established their own forms of representative government during the California Gold Rush. Drawing on the the early mining codes of medieval German miners in the Harz Mountains, on the mining customs of the Cornish tin miners and early Spanish mining laws introduced into California, the miners established the first governments in the American West. 340 ppgs, 24.99

BLM Field Handbook for Mineral Examiners - Leading mining historian Kerby Jackson introduces us to a classic mining publication on mine evaluation. First published in 1962, this work sheds important light on the techniques of BLM Mineral Examiners to perform validity on mining claims. 132 ppgs, 10.99

<u>**Six Months In The Gold Mines During The California Gold Rush**</u> - Unavailable since 1850, this important work is a first hand account of one "49'ers" personal experience during the great California Gold Rush, shedding important light on one of the most exciting periods in the history of not only California, but also the world. Compiled from journals written between 1847 and 1849 by E. Gould Buffum, a native of New York, "Six Months In The Gold Mines During The California Gold Rush" offers a rare look into the day to day lives of the people who came to California to work in her gold mines when the state was still a great frontier. **8.5" X 11", 290 ppgs. Retail Price: $19.99**

<u>**The Discovery of Gold in Australia**</u> - **First published in 1852, it has been unavailable since those days and sheds important light on Australia's gold mining history. Included are rare communications between British agents and the British Crown when gold was first discovered in Australia in 1851. This rare text contains hard to find details on Australia's first mining camps and Britain's early attempts to provide for the orderly regulation of gold mines in that part of the world. Also of interest are hard to find extracts of articles that appeared in the early colonial newspapers that did their best to report on Australia's gold rush as it took place.**
102 ppgs, 10.99